SHEEPWAGON
Home on the Range

August, 20

My father, Albert Parantean, was a sheep shearer in this area in early 1940's.
CP

SHEEPWAGON
Home on the Range

Nancy Weidel

HIGH PLAINS PRESS

9 8 7 6 5 4

Library of Congress Cataloging-in-Publication Data

Weidel, Nancy.
Sheepwagon : home on the range / Nancy Weidel.
p. cm.
Includes biographical references and index.
ISBN 978-0-931271-63-2 (hard) --
ISBN 978-0-931271-64-9 (paper)
1. Sheep herding--West (U.S.)--History.
2. Shepherds--Dwellings--West (U.S.)--History.
3. Wagons--West (U.S.)--History.
I. Title.

SF375.4.W+
636.'08'45--dc21 2001024269

HIGH PLAINS PRESS
539 CASSA ROAD
GLENDO, WY 82213
WWW.HIGHPLAINSPRESS.COM
ORDERS: 1-800-552-7819

To my parents, Daniel & Antionette Weidel

CONTENTS

Author Nancy Weidel is a historian and freelance writer. Raised in Pennsylvania and Delaware, Weidel received a master's degree in history from the University of Delaware.

She has lived and worked in South Dakota, Montana, and Oregon. She currently resides in Cheyenne, Wyoming, and has worked at the Wyoming State Historic Preservation Office since 1992.

ACKNOWLEDGMENTS

MANY PEOPLE HELPED ME with this book, beginning with a wonderful teacher, Dr. Bernie Herman at the University of Delaware, who inadvertently dropped the idea of sheepwagons in my lap. Next, good friends Gene and Kandi Davis, of Sheridan, Wyoming, who first saw this as a book when I thought it was merely an interesting research project for my spare time.

Wyoming sheep ranchers helped me a lot. I spent days in their pickups, desperately trying to keep a tape recorder going above the rattle of the range, as they conveyed their personal histories of the sheep business and love of the land. I could never have written such a book without their stories and invaluable information. Very special thanks go to ranchers Vern and Della Vivion, Leonard Hay, Bill Taliaferro, Don Meike, Simon and Dollie Iberlin, and John Etchepare, who so patiently endured endless stupid questions from a gal who grew up in an East Coast 1950s suburb.

Richard Collier, ace photographer at the Wyoming State Historic Preservation Office, helped immensely in the early stages of this project, accompanying me on various "sheepwagon road trips" and skillfully reproducing so many essential photographs. Also, thanks to John Keck, Director of Wyoming State Parks and Cultural Resources, who recognized the importance of the sheepwagon and sheep business to Wyoming's history.

Many thanks must go to the Wyoming Council for the Humanities and the Wyoming State Historical Society who early on granted funds that allowed me to independently pursue my study of sheepwagons; their support was invaluable. Also, Wyoming's Ucross Foundation provided an idyllic month and great food, affording me the opportunity to spend my days just writing. What a luxury.

I learned so much from so many, like former sheepherder and friend Louise Turk. Her generosity and kindness have been a gift. The dedicated sheepwagon restorers around the West who shared their technical knowledge with me—including Rawhide Johnson, Terry Baird, Gene Crosbie, Pete and Sandy Roussan, the late John Burke, and so many others—taught me so much.

A patient, flexible, and totally understanding publisher, Nancy Curtis, and a great editor, Mindy Keskinen, suited my haphazard style perfectly. How lucky I was they took me on for my first book.

Finally, thank you to my sister and lifelong best friend, Susan Weidel, for her unwavering support of my work. She provided invaluable assistance, superlative wit, great enthusiasm, and much-needed organizational skills; she even moved from Chicago to Wyoming to come to my aid (at least that's my version of her story).

Thank you to everyone!

P A R T   O N E

THE STORY

The Old Sheepwagon [1]

On a creek in old Montana
Fanned by a gentle breeze
Stands an old sheep wagon
Among the cottonwood trees

Its door is battered and sagging
Its cover is tattered and torn
Its stove is falling to pieces
And its wheels are warped and worn

It stands out in the weather
A relic of days that are past
This humble old sheepwagon
Is one of the very last

And the woolies that bedded around it
Have gone ere long ago
The herders are gone and forgotten
Some are sleeping beneath the snow

People pass it by unheeding
"A worthless wreck" they say
They do not seem to realize
That there has been a day

When all the world depended
On this range for wool and meat
That around this old sheep wagon
Once grazed a band of sheep

— E. F. Bischof

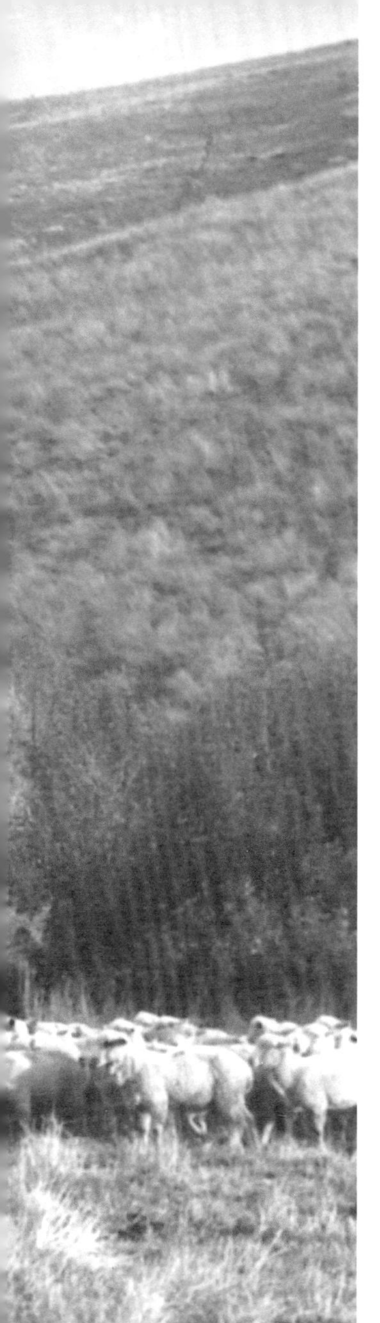

A sheepherder tends his flock on a Montana summer range. (Photo courtesy Museum of the Rockies, Montana State University)

C H A P T E R   O N E

The Early Days of the Western Sheep Business

THE STORY OF THE enormous sheep drives that took place in the American West over a hundred years ago has yet to be fully appreciated. So few images, so few reports of sheep drives were captured for the history books that it is almost as if they never happened. Cattle drives, yes. Cattle drives play a colorful role in American history. Portrayed in classic movies such as *Red River* and evocative novels like Larry McMurtry's *Lonesome Dove*, the western cattle drive has become a powerful myth, part of the American psyche. The drives are not hard to imagine when one stands on a gentle hill in the middle of a desolate landscape of the western Great Plains. Squint just a little and you can almost see the herds heading north out of Texas, west from Kansas, climbing slowly higher with every step as they cross onto the high plains of Wyoming and Montana. Close your eyes and you can hear them on the wind, just over the next rise.

But sheep drives? No films romanticize the historic sheep drives, no frontier photographer documented them, no books celebrate them. Nevertheless,

Wyoming sheepwagon, circa 1900. (Photo courtesy Wyoming Department of State Parks and Cultural Resources)

millions of sheep made the same type of journey as cattle did to reach the interior western United States. One reason these drives are virtually unknown today has to do with the myth of the West, the romance of the West, the lens through which popular western history has been filtered for our viewing pleasure.

Cowboys working cattle developed a rough-and-tough image that fit a mythological version of the American male character. Handling sheep involved a different, solitary way of life and a nurturing skill with the animals, which, when contrasted with the cowboy machismo, seemed almost feminine in nature. Whereas cowboys were viewed as brave and fearless, sheepherders were often looked down upon as pathetic, almost comical characters.

The history of western sheep ranching has been for the most part ignored. Yet it is a dramatic story, and much like cattle ranching, features heroic feats by men and animals played against the grand and often fierce natural elements of land, water, and weather.

The story of the sheep drives could be called "We Pointed Them East" as many herds were trailed from Oregon and California eastward to Idaho, Montana, and Wyoming. Their routes backtracked over the well-established pioneer roads: the old Oregon,

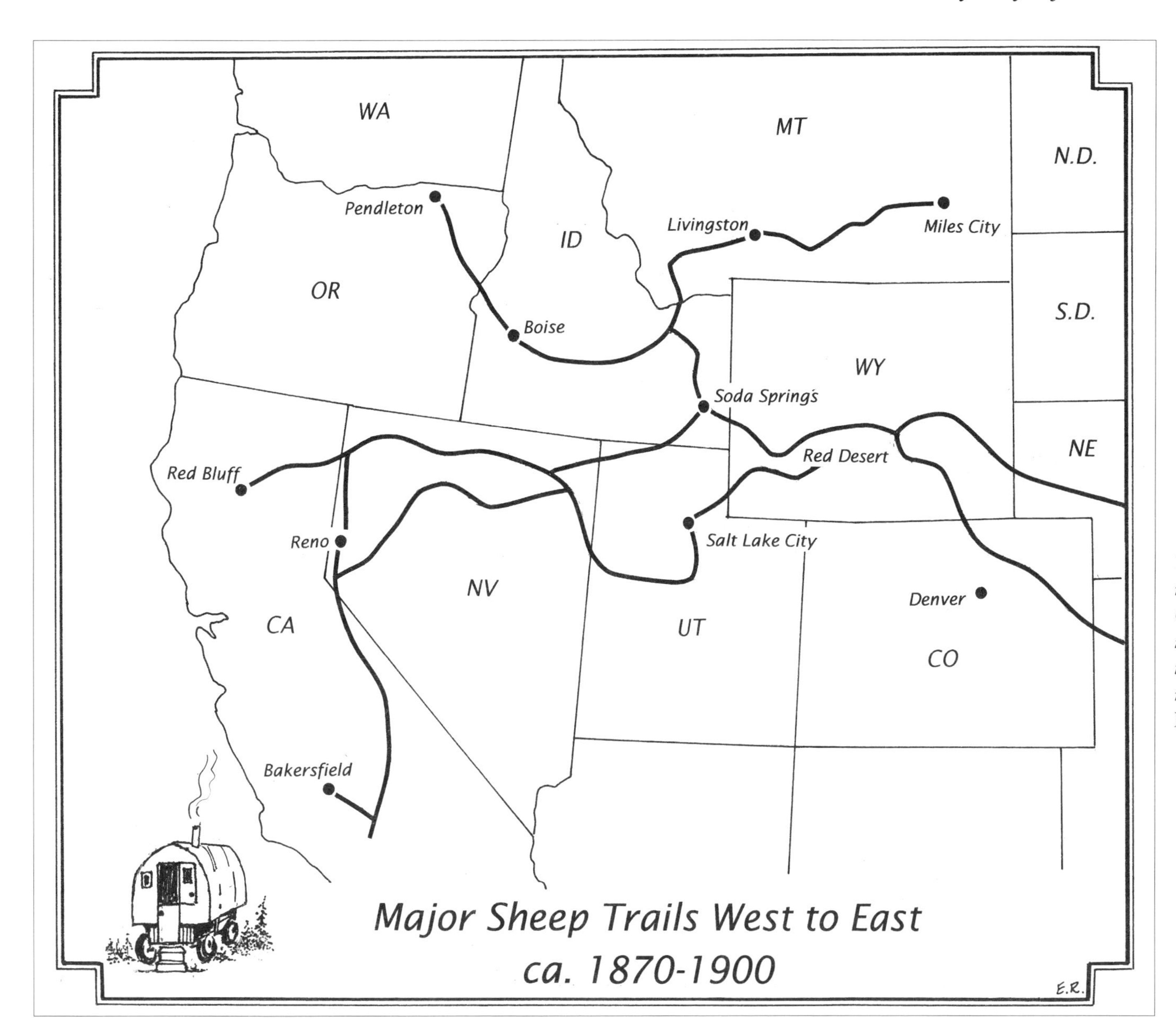

This map shows the major routes for trailing sheep from Oregon and California to the interior western states. These trails were used until approximately 1900. (Illustration by Elizabeth Rosenberg)

This photograph, taken near Douglas, Wyoming, shows a typical canvas top on a traditional sheepwagon. The grub box is open and accessible from the outside on this wagon. (Photo courtesy American Heritage Center, University of Wyoming)

Mormon, and California Trails. Typically, three men herded a band of between twenty-five hundred and five thousand sheep on a drive that could take as long as two years. The era of the great sheep drives lasted about thirty-five years, from 1865 to 1900, and it is estimated that at least fifteen million sheep made the eastward trek.[2]

Although cattle came first to the wide-open ranges of the West, sheep were not far behind. In Wyoming, the arrival of the Union Pacific Railroad in the late 1860s spurred the first permanent settlements by outsiders. Only twelve years later, by 1880, sheep numbers reached 500,000. By 1900, that number had increased over sevenfold to three and three-quarter million sheep.[3]

The devastating winter of 1886-87, which killed hundreds of thousands of cattle in Wyoming and Montana, actually helped the fledgling sheep business by freeing up large areas of the overgrazed range. As cattle entrepreneurs, who were often from wealthy English or Scottish families, fled the formerly profitable business (many returned to Europe), land became available for the sheep industry's expansion. The range wars of the next two decades between cattle and sheep ranchers testified to the growing power and prosperity of the sheep business by the late 1800s.

Due to arid conditions in much of the West, hundreds of square miles of grazing range, with a variety of grasses and sage, were absolutely essential to a successful cattle or sheep operation. As it still is today, grazing land was measured in sections, comprised of six hundred and forty acres each. Transhumance, a method of livestock management in which animals were trailed great distances in the yearly grazing cycle, was first introduced to the New World in the 1500s by the Spanish conquerors of Mexico and what is now the American Southwest. This range practice migrated north as the first large herds of cattle, trailed from Texas, moved into Montana and Wyoming in the 1870s.

Transhumance differed substantially from the stock raising methods of the eastern and midwestern United States, childhood home to many early cattle and sheep ranchers. The hundredth meridian divides these climatic zones. Imagine a straight north-south line that skirts the west bank of the Missouri River in the Dakotas, cuts the states of Nebraska and Kansas in half, and slices through eastern Oklahoma and Texas: this is the hundredth meridian. Areas to the east can receive several feet of yearly rainfall; the land west of the hundredth meridian receives on average less than twenty inches of moisture annually, the minimum needed for unirrigated agriculture.[4] A much more compact type of animal husbandry, consisting of farm flocks, could thus be accomplished in the East on smaller areas of well-watered land which produced lush grass. Feed such as hay, oats, and corn could also be easily grown during the long summers east of the hundredth meridian, not the case in the short, dry growing season of the West, where irrigation was needed to assure a decent crop.

Land ownership also played a crucial role in the two types of animal management. While eastern farmers owned their land, using it themselves or leasing it out, many square miles in the West were the property of the federal government or railroads such as the Union Pacific in Wyoming and the Northern Pacific through Montana. In the days before the establishment of such federal agencies as the National Forest Service and the Bureau of Land Management, public land had little oversight. The first-come, first-served principle operated to the advantage of the early cattle ranchers and companies, but plenty of open range remained for those who followed ten or twenty years later.

A sheepherder in the Big Horn Mountains summer range, northern Wyoming. Many herders enjoyed the summer months in the mountains. (Photo courtesy American Heritage Center, University of Wyoming)

Homestead parcels in the 1870s and 1880s consisted of as few as one hundred and sixty acres, which in the dry western states was enough land to graze only four cows or fifteen sheep. Free access to the vast public lands was crucial to the success of large ranching ventures, since only the wealthiest stockmen could purchase the land needed to feed hundreds of animals over a yearly grazing cycle.

In late spring and early summer, sheep moved to the mountain summer range in bands of nine hundred to a thousand ewes, along with their lambs. In places like southwestern Wyoming, the summer range could be a hundred miles from ranch headquarters, and trailing might take seven to nine days. The distance of a day's trail depended largely on water sources. Typically, the herder, starting mid-morning, trailed the sheep to water by noon where they rested for three or four hours during the hottest part of the day. The sheep then moved halfway to the next available water source, where they bedded down for the night.

Other herds trailed even farther to the summer range, taking as long as three weeks. In that case, the trail itself provided much of the actual summer grazing ground. As old-timers recalled, you might reach the top of the mountains just in time to turn around

The Hemry family at home on the winter range, Wyoming, 1902. (Photo courtesy, Casper College Library, Hemry Collection)

and head back down before the first fall snows arrived in the high country. So sheep depended on the fresh grass of the ascending and descending stock trails, as well as the mountaintops.

The vast winter ranges were located at lower elevations, but still remote from towns. Here the strong winds, so bitterly cursed in the West, became a friend to the sheepman. Although many inches of snow might fall, it often blew off the various grasses and sage that were staples of the range animal diet.

Sheep lambed on the range in the early days, with no shelter other than the landscape of ridges and gullies, or such vegetation as large clumps of sage or scrub brush. Following the birth of the spring lamb crop, crews of shearers removed the wool from the ewes and the wool was shipped to eastern markets, centered in Boston. Then the yearly cycle began anew with the annual trek to the summer range. Young lambs could safely take to the trail when they were about a month old.

Carbon County, Wyoming, blizzard of 1949. The National Guard airlifted feed for livestock stranded in the blizzard. (Photo courtesy Wyoming Department of State Parks and Cultural Resources)

Unlike cattle, sheep required humans to watch over and protect them from sudden storms, and from predators such as coyotes and mountain lions. The shepherd, or sheepherder as they are called in the West, served this function and played a vital role in the success of the sheep business.

In grazing areas of the Southwest, the herder and his dog followed the sheep with a minimum of gear. Dogs were always an important feature of the herding operation for not only did they assist in working the sheep, they were also the herder's sole companion during lonely weeks and months on the isolated range. One or two pack animals, usually horses, carried supplies, and the herder often slept on the ground beneath the stars.

Due to the harsh weather of the high plains and mountains of the northern states, however, herders there needed protection on the open range from the snow and winds of winter and the severe storms of summer. A shelter for the herder had to be mobile and portable so it could easily follow the sheep to the various grazing grounds. The herder's home also needed a source of heat for warmth and cooking. Invented and adapted for this specific purpose, the sheepwagon became the ideal home for the herder.

PART TWO

THE WAGON

Memories of Yesterday[1]

Out on the hilltop
all weathered and bleak
stands an old sheepwagon
the symbol of sheep.

The tongue is a wreck
there's holes in the floor
and the wheels will turn
as they did, nevermore.

From where I now sit
all wrinkled and gray
fond memories return
of yesterday

I can still hear the bleating
on the evening bed ground
there's no duplicating
that musical sound.

Melancholy by sunshine
lugubrious by rain
she withstood blizzards
droughts, misuse and strains.

You could see her white cover
for many a mile
and the comfort she offered
brought many a smile.

Now we're both battered
we're old and we're gray
I want to rebuild her
as she was yesterday.

— Bill Norton

THE DENVER POST
U.S. COURT UPHOLDS
BULLETINS!
MAJOR IS
SENTENCE
STAPLETON

C H A P T E R   T W O

Inside the Basic Sheepwagon

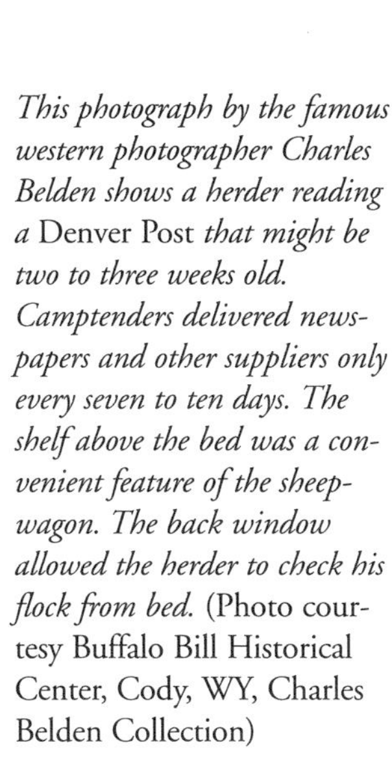

This photograph by the famous western photographer Charles Belden shows a herder reading a Denver Post *that might be two to three weeks old. Camptenders delivered newspapers and other suppliers only every seven to ten days. The shelf above the bed was a convenient feature of the sheepwagon. The back window allowed the herder to check his flock from bed.* (Photo courtesy Buffalo Bill Historical Center, Cody, WY, Charles Belden Collection)

THE SHEEPWAGON IS A marvel of practicality and efficiency. The traditional design of the sheepwagon, along with the placement of such defining interior features as bed, stove, table, and benches, had become standardized by 1900, just sixteen years after its invention. But that does not mean every sheepwagon was built just alike; in fact, quite the opposite is true. One can think of a sheepwagon as standardized in the sense that today's automobile is. Various chassis types are designed to hold different-sized body frames, engines contrast in size and power, and both the interior and exterior can be customized with a myriad of accessories. It is impossible to describe a "standard" car and the same is true of a "standard" sheepwagon. But as with the car, the sheepwagon's specifications fall within a certain range, so that it is possible to characterize a typical circa-1900 sheepwagon.

Designed to provide shelter and heat, mobility and storage, the sheepwagon was the ideal home for the herder. Approximately eleven and a half feet long and six and a half feet wide, enclosed by a canvas top,

This wagon has a movable tailgate on the back, rather than a grain box. The gate held equipment such as saddles and tools. (Author's photo)

it held a wood- or coal-burning stove. It could easily be moved by two horses, a most important feature. The wagon contained plenty of storage areas, both inside and out. A large box built on the back of some wagons held grain; other wagons had a movable tailgate in this same spot which carried equipment. Wagons often had smaller exterior mess boxes built on the sides of the wagon box, between the front and back wheels; some of these could be accessed from both inside and outside the wagon. The side boxes also came in handy during lambing, when a weak newborn might be placed there overnight to be revived by the heat of the wagon stove.

The door was at the front, where the herder stood to drive the wagon when underway. Most doors were six feet high, a few inches shorter than the interior height of the wagon. As one stepped into the wagon from the tongue, a stove stood to the immediate right with the stovepipe extended through a hole in the canvas top. Some type of built-in cupboard or shelf usually occupied the space behind the stove. Easy-to-clean oilcloth often covered a portion of the canvas wall and ceiling near the stove. Each long side of the wagon had a bench for seating that doubled as storage space; a hinged lid provided access

No two sheepwagons were identical. These two photos show the back of the wagon's interior with the bed and table, as seen from the front door. Note the difference in the tables. The table at the left is supported by a gate-leg, while the table below is cantilevered. The wagon at the left has only four drawers under the bed, while the wagon below has both a cupboard and drawers for storage. The interior of the wagon at left is varnished, while the wagon interior below is painted. (Left photo courtesy Richard Collier, Wyoming State Historic Preservation Office, Department of State Parks and Cultural Resources; lower photo by author)

This Charles Belden photograph shows the Dutch door at the front of the wagon with the bottom closed and the top portion open. Note the top door also contains a window. Stepping onto the wagon tongue permitted easy access to the interior. The herder is seated on a water barrel. (Photo courtesy Buffalo Bill Historical Center, Cody, WY, Charles Belden Collection)

to the storage space within the bench, which held flour, meat, or other food supplies. Some wagons had a built-in flour box on the end of the bench across from the stove, just left of the front door. The water jug usually rested in this corner, on the bench.

Opposite the front door, the bed platform stood about three feet high, perpendicular to the side benches. An operable window was always built in the back wall of the wagon, above the bed. A table, actually a long slab of wood, slid out from a slot under the bed and could be either cantilevered, supported by a gate leg, or suspended by a chain hung from the ceiling. With the table extended for eating, the side benches provided comfortable seating for the diners. The table also came in handy as an extra bed; it could be pulled all the way out from its cavity, with each end resting on a side bench to create a sleeping surface.

The storage space under the bed platform and table slot was accessed either by doors, drawers, or a combination of both, depending on the wish of the buyer or the whim of the builder. Sometimes the builder left the space open so a herder could store bulkier supplies such as a saddle, lamp, kerosene, and boots.

Although the first wagons had only a canvas flap for a door, a Dutch door, or "stable door" as the English called it, quickly replaced the flap and became one of the sheepwagon's most prominent features. The door's top half could remain open while the bottom stayed shut. This functional feature served several purposes; with the top open, the herder could hear and see his sheep. The open top door also provided ventilation for the wagon and modulated the heat of the stove, which could be quite intense. A closed bottom door also kept out the herder's dogs and other animals.

According to ranchers, however, the primary function of the Dutch door was to allow a herder or camptender to stand within the wagon—or even sit on the side bench or a box—and still be able to extend his arms through the open top door to hold the horses' reins when the wagon was being moved. The bottom door, the smaller of the two, remained shut during a move to protect the driver. From within the wagon, the driver could also operate the brake, usually located to the right as he faced out.[2]

Most sheepwagon restorers will tell you that no two sheepwagons are exactly the same, a fact easy enough to verify if one takes a tape measure to them. Sheepwagons ranged in length from ten to twelve feet; they could be six and a half feet wide or closer

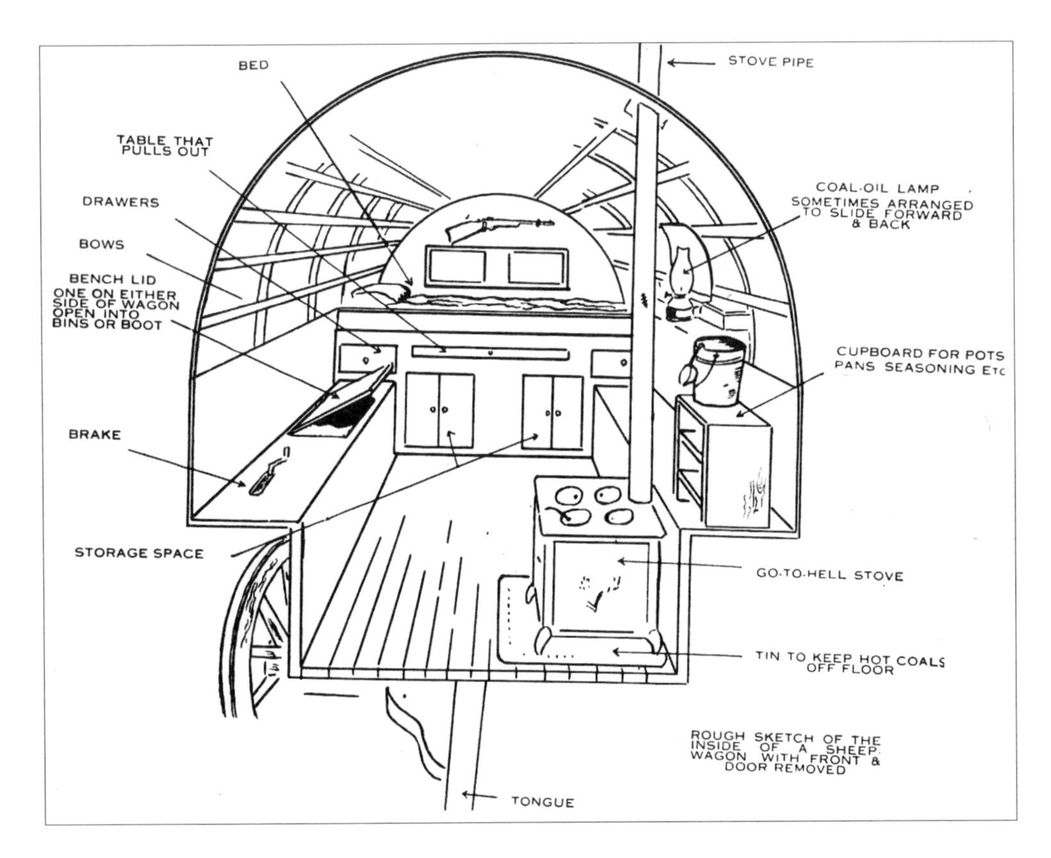

Illustration of the interior of a sheepwagon. Almost every sheepwagon followed this basic design with individual variations. (Source: *Ten Sleep and No Rest: A Historical Account of the Range War of the Big Horns in Wyoming* by Jack R. Gage, Prairie Publishing, Casper, Wyoming, 1958. Reprinted with permission.)

to seven feet. One builder used only four bows, another as many as eight.

Inside the wagon, the variations were even greater. The window above the bed might be a three-paned slider or a single-light hopper that latched to the window frame when in the open position. Many wagons had two windows, others only the rear one. Some tables were small and folded against a side wall when not in use. The standard bed was three-quarter sized but a full-width bed was a feature of other wagons. A variety of manufactured stoves were used in the wagons.

Remarkably, the interior configuration of the sheepwagon proved so efficient that one hundred and sixteen years after its invention, the same basic plan is used in the few new wagons made today. Whether by design or accident, the sheepwagon interior also served as the model for many modern campers. One can begin to appreciate the durability of the sheepwagon's simple interior design in contrast to the many variables of home floor plans. The sheepwagon has retained its original interior configuration because the placement of its door on the front rather than the back, and its component parts such as the bed and stove, utilized the small space in the most efficient manner possible.

This wagon, missing its canvas top, shows the structure of a traditional wagon. This wagon has six bows. The number of bows on a wagon ranged from four to eight. (Photo courtesy Martha Gibbs collection)

CHAPTER THREE

The Origins of the Sheepwagon

Moving a wagon downhill was always a precarious job. These two men (probably the herder and the camptender) are struggling to control the downhill momentum, circa 1908. (Photo courtesy Mary Vawter Duggleby collection)

THE SHEEPWAGON IS AN object of fascination to many people today. Like other commonplace articles from our past that have slowly slipped into obsolescence, the sheepwagon has become a symbol of a seemingly purer, less complicated way of life. The sheepwagon conjures up an image of a self-sufficient man alone, a simple caretaker of helpless animals in the solitude of pristine mountains and desert, a life reduced to its most elemental, far from the polluting influence of civilization. It seems the further such a romanticized past recedes, the closer it slips toward the vanishing point, the more quaint and interesting it suddenly becomes.

So it is with the sheepwagon. People often wonder who invented the sheepwagon, speculating that the first ones were modeled on the Conestoga wagon. Actually, the genuine Conestoga wagon, manufactured in southeastern Pennsylvania beginning in the 1750s, was most heavily used between 1820 and 1840, and probably few made it west of the Mississippi River.[1]

The Conestoga wagon is often confused with the "prairie schooner" that transported many pioneers across the country. The prairie schooner, however, was primarily a freight vehicle, a type of moving van used to carry the pioneers' supplies and possessions to a distant new home. Unlike the sheepwagon, the schooner was not outfitted with a bed and stove, and conveyed only the elderly, young children, or the ill on the journey westward. In fact, the healthy pioneer spent much more time on foot than inside a wagon on the westward journey.

Others have noted similarities between a type of nineteenth-century military vehicle, the ambulance, and the sheepwagon. Although the Army ambulance had its own military purposes, such passenger vehicles as the buckboard, the mountain wagon, and especially the Dougherty spring wagon were often termed ambulances on the western frontier. An Arizona military wife described such a wagon: "A Dougherty wagon or in common army parlance an ambulance was secured for me to travel in. This vehicle had a large body, with two seats facing each other and a seat outside for the driver. The inside of the wagon could be closed if desired, by canvas sides and back which rolled up and down by a curtain which dropped behind the driver's seat." [2]

Likewise, Elizabeth Custer, wife of the famed General George Armstrong Custer, described the ambulance her husband had fitted up as a traveling wagon for her as "quite a complete house in itself." [3]

By far the most harrowing account of a journey in a modified-for-long-distance-travel Army ambulance is to be found in the book *My Army Life and the Fort Phil Kearny Massacre*, written by the wife of Colonel Henry Carrington. Colonel Carrington commanded the beleaguered Fort Phil Kearny located near present-day Story, Wyoming. Mrs. Frances Carrington described a trip from the fort to Omaha, Nebraska in January 1867, one month after the infamous Fetterman Massacre. Women and children made this perilous journey, through hostile Indian territory during a raging blizzard, in a wagon she described in some detail and called "my traveling house."

General George B. Dandy fitted up the army wagons in which they were to travel "in a manner that proved our very salvation on such a journey...It was a novel caravan indeed. The wagon covers of cloth were first doubled, and both sides and ends of the wagon bodies were boarded up, with a window in each end. A door at the back of each wagon swung on hinges to admit of easy ingress and egress, and near the door was a small sheet iron stove made from stove-pipe, with a carefully adjusted smoke escape through the wagon-cover above...The best modern kitchen-range could not secure as far reaching results as our little sheet-iron stoves." [4]

The interior of the western sheepwagon also resembles a ship's cabin, another version of a compact, efficient living space. The similarities are remarkable: both have well-designed storage areas, built-in benches, retractable or fold-down tables, and a sleeping berth. A story is told of two old sheepherders, both former sailors, who finally felt at home in the western desert which they described "like being on a dry sea." A term often applied to the deserts and high plains is "a sea of grass," the empty landscape consisting of only two elements, treeless earth against a huge sky, the undulating hills resembling waves on the ocean.

The idea of a covered wagon as shelter for a shepherd was not a new one. A relative of the sheepwagon was the four-wheeled "shepherd's hut" used in the British Isles during the nineteenth century and described by Thomas Hardy in his 1874 novel *Far From the Madding Crowd.*

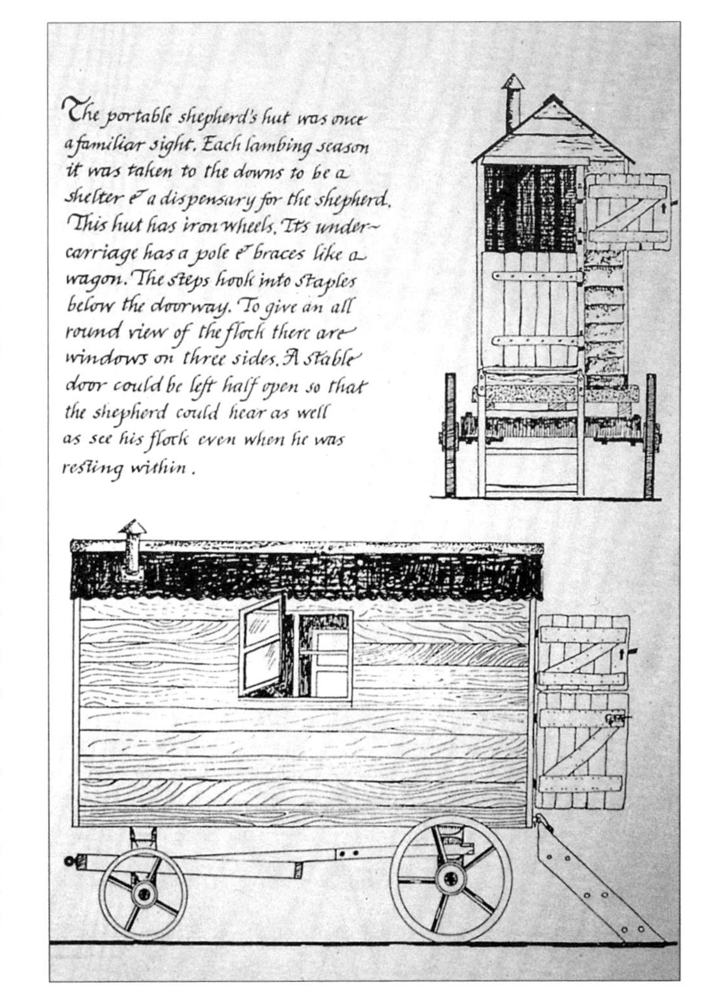

Shepherd's hut. This wagon was used in England primarily during lambing season. Unlike the sheepwagon, it did not house the shepherd full time and did not travel great distances. Similarities to the sheepwagon include the Dutch door, windows, stovepipe and wooden running gear. (Source: *Old Farms, An Illustrated Guide*, John Vince, Schocken Books, NY 1983.)

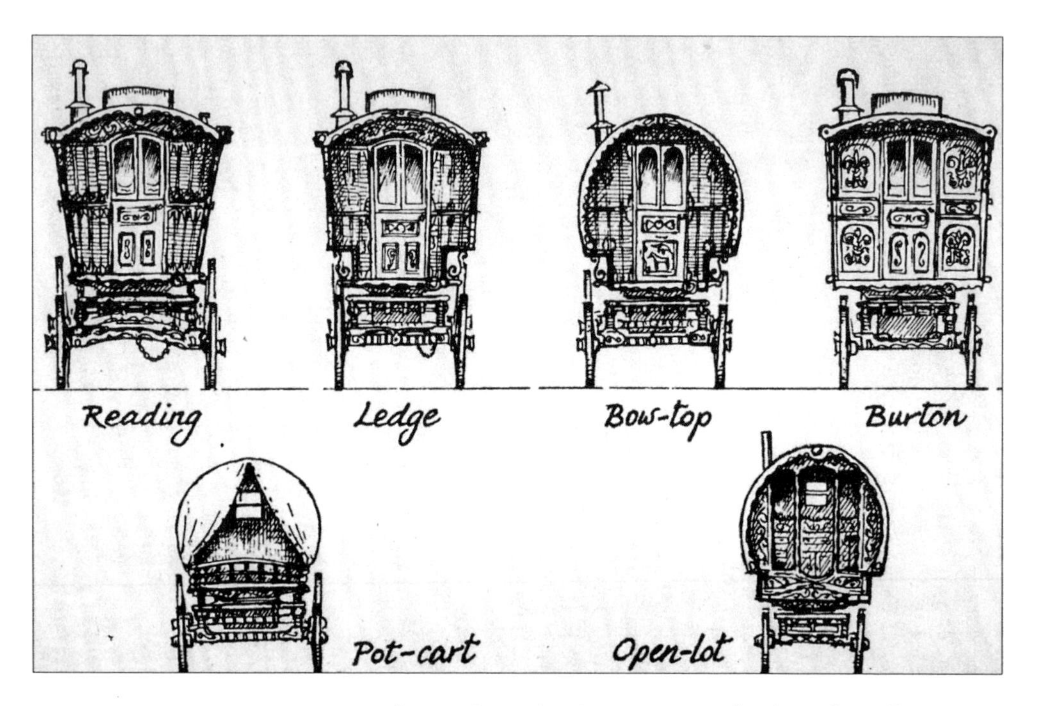

Above and at right: Gypsy Wagons. The photo above illustrates the various types of gypsy wagons. The sheepwagon resembles the bow-top. The photo on the opposite page shows a typical gypsy bow-top wagon, circa early 1900s. (Source: *The English Gypsy Caravan: Its Origins, Builders, Technology and Conservation,* C.J. Ward-Jackson and Denis E. Harvey, Drake Publishers Inc., New York, 1973.)

"The hut stood on little wheels, which raised its floor about a foot from the ground. Such shepherds' huts are dragged into the fields when the lambing season comes on, to shelter the shepherd in his enforced night attendance. The hut had a door, a small stove, and the bed which consisted of a rather hard couch, formed of a few corn sacks…covered half the floor of this little habitation. In the corner stood the sheep-crook, and along a shelf at one side were ranged bottles and canisters...On a triangular shelf across the corner stood bread, bacon, cheese...The house was ventilated by two round holes, like the lights of a ship's cabin, with wood slides." [5]

The sheepwagon as we know it today is clearly related to Hardy's hut catering to the needs of the shepherd. But the shepherd's hut was a smaller, heavier wagon, not designed to travel hundreds of yearly miles over the rough terrain of the American West.

The most direct antecedent to the western sheep-wagon is the traveling gypsy van or wagon. This house on wheels appears to have originated in France; it became popular in the nineteenth-century British Isles as the road system improved. By the 1830s, the gypsy wagon provided a mode of conveyance and a movable house for a variety of non-gypsy people:

traveling showmen, itinerant tinkers, basket and brush makers, and peddlers.[6]

In spite of the wagon's generally accepted name, true gypsies, with their conservative, peripatetic ways, were slow to trade in their tents for the house on wheels, and did not begin to make the switch until after 1850. Thereafter, any traveling van, whether used by non-gypsies or by Romanies, as the authentic gypsies were sometimes called, became known as a gypsy van, an abbreviation of caravan or wagon.[7]

Living in a wagon in a less-than-temperate climate was made possible by the design of a portable kitchen furnace in the late eighteenth century. By the 1860s, a type of free-standing, cast-iron cooking stove called an "American stove," first marketed in the United States in 1830, came into common use in the British Isles.[8]

The conventional layout of the gypsy wagon was established by the early 1870s. Typically it measured larger than the traditional sheepwagon, and consisted of one room, twelve feet long by six feet wide and six feet high. A front Dutch door provided a more efficient use of space than a side door. Standard interior features of the gypsy wagon, like the sheepwagon, included a bed against the back wall, built-in

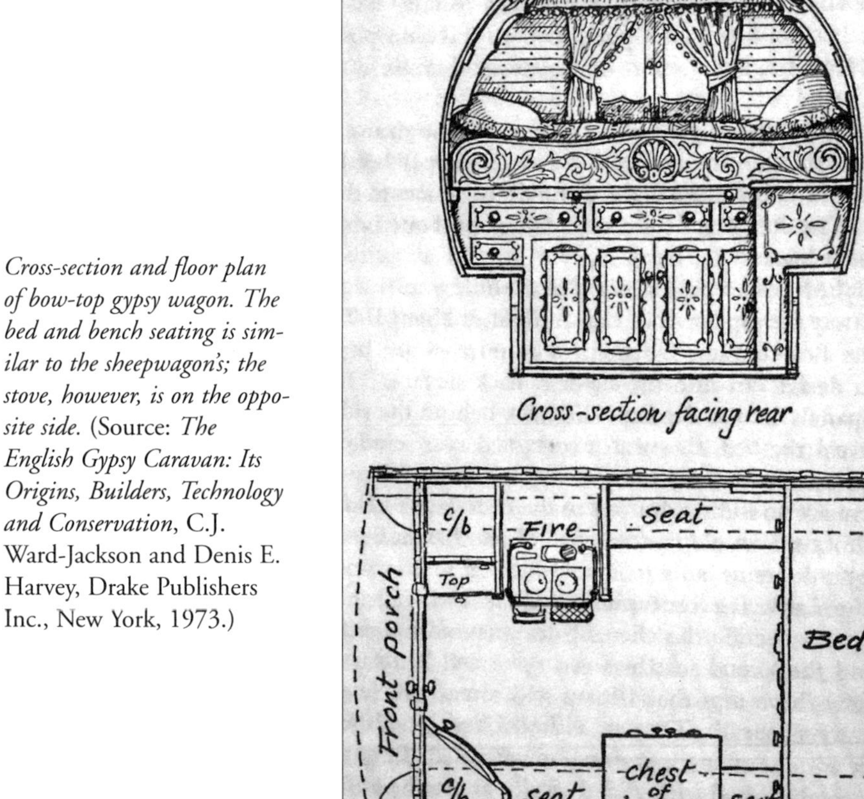

Cross-section and floor plan of bow-top gypsy wagon. The bed and bench seating is similar to the sheepwagon's; the stove, however, is on the opposite side. (Source: *The English Gypsy Caravan: Its Origins, Builders, Technology and Conservation,* C.J. Ward-Jackson and Denis E. Harvey, Drake Publishers Inc., New York, 1973.)

storage compartments, and the essential stove. Although the gypsy wagon could be quite fancy, with interior details such as carved mahogany cabinets and elaborate mirrors, there is no doubt it is a relative of the western sheepwagon. A photo of a particular type of gypsy van, known as the bow-top, confirms the remarkable resemblance between the two.

Descriptions of gypsy wagons appear in literature. In Kenneth Grahame's childhood classic, *The Wind in the Willows*, Toad is eager to begin a life of wandering and proudly shows off his new "gipsy caravan" to his friends Mole and Rat.

"It was indeed very compact and comfortable. Little sleeping-bunks—a little table that folded up against the wall—a cooking-stove, lockers, bookshelves, a bird-cage with a bird in it; and pots, pans, jugs and kettles of every size and variety.

'All complete!' said the Toad triumphantly, pulling open a locker. 'You see—biscuits, potted lobster, sardines—everything you can possibly want... you'll find that nothing whatever has been forgotten...'" [9]

Prominent western sheepman Robert Taylor, a native of Scotland, compared the sheepwagon and the gypsy van in a letter to his brother written from

Much like the sheepwagon, the gypsy wagon is being restored and displayed. This photo was taken in Brighton, England in 1996. (Photo courtesy Richard Collier)

Wyoming in 1886: "I am writing this in one of my sheep wagons. These are filled like a cheap traveling van in the old country with stove, bunk, etc. We move around all winter." [10]

Although no documentary evidence appears to exist that would confirm it, most experts agree the first sheepwagons were probably improvised by sheepmen of the northern territories since it was here that they needed protection from the harsh weather. They began with a standard wagon box about ten feet long and three feet wide. The bows of such light freight or farm wagons could easily be covered with canvas and outfitted with perhaps a bedroll inside. Adding a built-in bunk and fastening a small stove to the floor were the next logical steps in the evolution of the crude house on wheels. [11]

As the sheep business rapidly expanded throughout the West in the 1870s and 1880s, and transhumance became established as the most efficient method of production, demand increased for a type of portable shelter capable of housing a man or two while moving sheep over thousands of acres of vacant landscape. Although sheepmen continued to make their own wagons, blacksmiths also began to build and no doubt improve upon them, selling their models at a profit.

James Candlish, the "inventor" of the sheepwagon. (Photo courtesy Carbon County Museum, Rawlins, WY)

Various men have received credit for the invention of the sheepwagon including Utah sheepman William McIntosh in 1880[12] and Wyoming blacksmith Jacob Jacobsen in 1883.[13] But the honor generally goes to Wyoming blacksmith James Candlish for the first wagon he built in 1884. Candlish (according to some sources in collaboration with local sheep rancher George Ferris) apparently took the four-wheeled, canvas-covered farm wagon a step or two closer to what would become the standard sheepwagon form.[14] Although his source material was not identified, author Clel E. Georgetta described the Candlish wagon of 1884 in his book *Golden Fleece in Nevada:*

> *He took an old lightweight wagon running gear he had behind his shop. He put on it an ordinary wagon box with high sideboards. Along each side he put on top of the sideboards a plank extending out over the top of the wheels. On top of the planks at the rear end he built a bunk bed about seven feet from head to foot and as wide as an ordinary double bed. Over the top of all he put bows and stretched canvas like the old "covered prairie schooner." To protect the*

James Candlish in his blacksmith shop in Rawlins, Wyoming. (Photo courtesy Wyoming Department of State Parks and Cultural Resources)

The far right wagon is a sheepwagon, circa 1900 at Walcott, Wyoming. This wagon has an interior canvas flap that appears to have been used as a door. The stovepipe is in an unusual position in the rear of the wagon which indicates the bed is not in its normal spot. (Photo courtesy Wyoming Department of State Parks and Cultural Resources)

herder from the wind on cold winter nights he put, just in front of the bed, a canvas partition parted in the middle like flaps of a tent. In front of that he built a cupboard that sat on one of the extending side planks and with the back fastened to the top bows. Near the front end of the wagon, upon the extending plank on the left side, he put a small sheet iron stove with the stovepipe extending up through the canvas top where he covered the canvas with tin—top and bottom—so it would not burn.

The front end of the Candlish "Home on Wheels" was all open, and to get into the rig, a herder had to climb up over the front endgate of the wagon. There was no door. On the outside, between the front and rear wheels, he put a small platform for the water keg.[15]

Apparently no photo exists that illustrates the interior of a genuine Candlish wagon, and Georgetta does not mention benches although other writers claimed the first wagon included them. However, an exterior photo of a Candlish model shows a canvas flap rolled

Commemorative fob honoring sheepwagon inventor James Candlish. The Wyoming Wool Growers Association gave a fob to each member in 1909. (Photo courtesy Richard Collier, from display at Laramie Plains Museum in Wyoming.)

up and secured above the entryway. A stovepipe and one exterior supply box can be also seen in this undated, but no doubt prior to 1900, photograph.

Both Candlish and rancher Ferris lived in Rawlins, a small town located on the first transcontinental railroad, the Union Pacific. By the 1880s, sheep were becoming big business in the town due in large part to its key location along the tracks which enabled the wool and meat products to reach markets back east in a timely fashion.

The Wyoming Wool Growers Association gave credit to Candlish as the inventor: in 1909 it commissioned a commemorative fob made of three small discs linked by a chain. The top disc noted the "5th Annual Convention Wyoming Wool Growers Assn. January 11-12, 1909." The middle disc bore a likeness of "The Inventor James Candlish." The bottom disc featured a sheepwagon and proclaimed "The Modern Sheep Palace Was Made in Rawlins, Wyoming 1884." Five hundred fobs were manufactured by Whitehall and Hoagg, a Kansas City firm, and each Wool Grower and his wife received one at the annual banquet that year in Rawlins.

Noted Wyoming author Agnes Wright Spring researched the invention of the sheepwagon for an

This early Candlish sheepwagon was allegedly built for Wyoming sheep rancher George Ferris. The first sheepwagons had canvas rather than wooden doors. The small tent, called a herder's tepee, was often used in the high mountain summer ranges. (Photo courtesy Carbon County Museum, Rawlins, WY)

article she wrote in 1940, possibly the first in-depth piece on the subject and a major source for many that followed over the next thirty years.[16] In response to her inquiries, Wright received a letter from one of James Candlish's daughters, Mrs. Fred W. Dodge of Ontario, Oregon, who supplied details of Candlish's life. Born in Montreal, Canada, her father learned the wagon making and blacksmith trades there. He made his way to the United States where he worked as a blacksmith for the Union Pacific at Omaha, Nebraska.

Candlish followed the building of the tracks west to Rawlins, then worked for the government at Fort Steele. He had moved to Rawlins by 1881 and established a blacksmith shop at the corner of Fifth and Buffalo Streets, near the site of the present-day Carbon County Courthouse. He remained in Rawlins until the early 1890s, then relocated to Lander, Wyoming where he died in 1911.[17] Mrs. Dodge credited her father with the design and building of the first sheepwagon which she referred to as the "house on wheels."[18]

The prominent Ferris family of Rawlins, however, disputed the claim that Candlish invented the sheepwagon independent of assistance from others. In a letter to the Wyoming State Historical Board in 1955, Mrs. Frank Ferris, daughter-in-law of sheep rancher George Ferris, wrote that "credit was given, and he accepted it, to Ed Cowdlish [*sic*] for inventing and making the first sheep wagon. The truth is this."

Mrs. Ferris explained that George Ferris came to Rawlins before the Union Pacific arrived in 1868, and became the second person to raise sheep there. She claimed that George Ferris designed and built the first sheepwagon which was "some larger than the present day wagon and a canvass curtain was used instead of a door." She added that in 1890, Ferris hired Ed Coudlish [*sic*] to build a sheepwagon "to his specifications" of which she had a picture. In closing she noted that her husband, Frank Ferris, was "nearing 80" and requested that "this little bit of history should be corrected while he is living."[19]

It is certainly possible that rancher Ferris did have a hand in the "invention" of the sheepwagon along with James Candlish. A collaboration between a sheep rancher and a blacksmith to develop a new type of wagon for living on the range makes sense. The blacksmith knew how to modify a wagon for a specific use, and the rancher understood how the wagon needed to function most efficiently for both living and working.

Restored Florence Hardware wagon, circa 1900, is on display at the Jim Gatchell Museum in Buffalo, Wyoming. (Author's photo)

While it's clear that Candlish played a role in modifying the standard wagon box into a sheepwagon, it is also apparent that other westerners were simultaneously developing the idea of an enclosed wagon to house a herder. In 1886, the *Douglas Budget*, a Wyoming newspaper, ran a small article below the headline "Sheep wagons dot expanse." The writer noted that "Florence Hardware Company of this town builds them and Manager Knittle has made many improvements during the *two years he has been building wagons*" (author's italics).[20]

A close runner-up for the title "inventor of the sheepwagon" was blacksmith Frank George of Douglas, most likely the builder of those wagons in the 1886 newspaper report. Born in Wisconsin in 1856, George trained as a wagonmaker. He arrived in Wyoming in 1877 and spent three years at Fort Fetterman repairing wagons with a Swedish wagonmaker named Charles Hogson, and Fred Ericson, a Norwegian blacksmith. George worked on wagons from Fort Reno and Fort Laramie and also did some freighting in the area.

Frank George moved from Fetterman to Douglas in 1887, shortly after the Fremont, Elkhorn & Missouri Valley Railroad reached that town. There he opened a wagon and blacksmith shop, and later worked for the Florence Hardware and Lumber Company. He is credited with building, according to various sources, either "the first sheep wagon ever built in Douglas,"[21] or "the first sheep wagons ever built,"[22] sometime in the late 1880s or early 1890s. The George wagon became known as the Florence wagon to identify the hardware store that was named after the wife of its owner, Mr. Knittle. A number of sheepwagons made by Frank George survive today; one is displayed at the Rockpile Museum in Gillette,

A Sixteen Horse String Team, Freighting Supplies to the Mountains

NOTHING LIKE WYOMING

Talk not to me of Eastern States, their cities large and grand;
With operas and seaside joys way down by Jersey's sand;
Of cafes fine and swell resorts, and functions up in G,
Wyoming, perched up near the sky, is good enough for me.

I've hit the trail to Iowa, and examined Kansas, too;
I've touched the highest spots they've got in far-famed old Mizzoo;
I've wintered in Los Angeles, also in Washington, D. C.;
But none of these can get my game—they're all too slow for me.

A. K. YERKES, Sour Dough Creek, Wyo.

This photograph with a two-wheeled cooster wagon was used to illustrate a poem celebrating Wyoming. This cooster also lacks a stovepipe which means the wagon did not contain a stove. (Photo courtesy Wyoming Department of State Parks and Cultural Resources.)

Wyoming. George also built the Florence Hardware sheepwagon now in the Jim Gatchell Museum collection in Buffalo, Wyoming.

It is worth noting that both James Candlish and Frank George were employed as blacksmiths at military forts in Wyoming Territory during the same decade, the 1870s. Obviously, they would have seen a variety of Army and freight wagons and most likely helped to winterize them, a routine procedure for the protection of military men and cargo against the fierce blizzards of the western plains. Author and wagon expert Richard Dunlop explained this process: "...the wagon ends were built up with boards. A hinged door was set in the back. The wagon covers were doubled, and a small sheet-iron stove was put inside with a pipe peaking through the tarp."[23]

The military has often been on the forefront of new technology and all types of vehicle adaptations. (One need only think of the army jeep, developed as the first four-wheel drive vehicle in World War II.) So it is not surprising that two talented army-employed blacksmiths would adapt military practices for a new type of civilian use. Coincidentally, both Candlish and George spent the 1880s in towns growing rapidly in part due to the rise of the local sheep business. It is therefore not impossible that both men independently conceived the idea of the sheepwagon at roughly the same time.

Various sources credit wagonmaker Marshall Buxton with further refinements to the sheepwagon, primarily on the interior, around 1900.[24] Employed by the Schulte Hardware Company of Casper, Wyoming, Buxton reportedly replaced the canvas front flap with the Dutch door (although one source reports that the bottom door was wood while the top was canvas and slid on a track, thereby eliminating the need for the canvas curtain in front of the bed). He also added such details as a shelf and operable window above the bed and installed a larger stove that included an oven.

Buxton also placed the sheepwagon on a type of running gear called "mountain gear," manufactured by the Bain Wagon Company of Kenosha, Wisconsin. The Bain wagon chassis was lighter than the standard farm wagon or freight wagon. It also had higher wheels and narrower tires than other running gears, making it better suited for traveling rough country and steep trails. The original Buxton/Schulte wagon cost $243, not including the Bain running gear which varied from $65 to $195.[25]

No doubt there are other claimants for the title of sheepwagon inventor throughout the West but it seems that credit must be shared. Research of patent records between 1870 and 1900 revealed that no patent was filed for a sheepwagon, house on wheels, or "sheep palace" as it was variously called, probably because the evolutionary idea could not be claimed by just one man at some specific time.

But by 1884 the sheepwagon had assumed its classic form. Sheep were arriving in the western states in large numbers, thousands being trailed yearly from California and Oregon throughout the 1880s. During that decade, a large sheep industry developed in such states as Wyoming, especially in towns like Rawlins, strategically located on the Union Pacific main line. The sheep industry in Wyoming tripled in size from about a half million sheep in 1880 to almost a million and a half a decade later.[26]

Therefore, although it is doubtful the sheepwagon was "invented" by one or two people, such men as Wyoming blacksmiths James Candlish and Frank George, and wagonmaker Marshall Buxton, certainly deserve credit for their design modifications. And so does every builder, however anonymous, who left his mark on each wagon he built.

James Julius "Cooster" Svendsen, holding the reins, may have invented the cooster wagon. (Photo courtesy Connie Norwood collection)

The Cooster Wagon: First Cousin of the Sheepwagon

The cooster wagon, a two-wheeled, canvas-covered wagon with a stove and a bed inside, bears a remarkable resemblance to the traditional sheepwagon. Freighters used the cooster wagon on their long hauls from remote settlements to the rail lines, hitching it to the rear of a cargo wagon. Like the sheepwagon, prototypes for the cooster wagon can be found in an early type of European gypsy wagon called a two-wheeled pot cart. This small canvas-covered, bow-topped cart was not meant to be lived in but did provide the traveling tradesman a place to sleep. There were many variants of the pot cart; a four-wheeled cart developed later gradually replaced the two-wheeled version. One author suggested that the two-wheeled cart was probably the "oldest kind of wheeled conveyance used by travelers."[27]

According to a 1991 article in the magazine *Western Horseman*, a Danish immigrant to Wyoming named James Julius Svendsen designed and built the first cooster wagon sometime in the latter part of the nineteenth century. Svendsen, who went by the nickname Cooster, apprenticed as a wheelwright in his native land and arrived in Wyoming sometime in the 1870s. A stint with freighters, who commonly camped out in the open, led him to build the first cooster. Svendsen moved to Casper in 1894, a time of widespread expansion in the nearby sheep business, where he adapted his two-wheeled cooster into a four-wheeled sheepwagon.[28]

Occasionally one finds a reference to a cooster wagon in a book. Perhaps "Cooster" Svendsen built the wagons that author J. T. Wall referred to when he wrote about hauling oil from a field to a refinery near Casper circa 1917 "... by long string teams consisting of sixteen to twenty horses, three oil-tank wagons and a cooster behind. The cooster was a small covered wagon that was used for the jerkline skinners to sleep in at night and to cook their meals."[29]

In his 1999 book *Echoes from the West*, Edward M. McGough described a cooster wagon. "The other two freight wagons and the cooster wagon were coupled

Freighters used a cooster wagon which was similar to a sheepwagon. This cooster wagon had two wheels instead of four. It was used by freighter George Coleman in 1898 in Wyoming's Big Horn Basin. (Photo courtesy Wyoming Department of State Parks and Cultural Resources, Coleman Collection)

A freight team with a cooster wagon in the rear in Lander, Wyoming, circa 1900. This photo was taken by noted Wyoming photographer Joseph E. Stimson.(Photo courtesy Wyoming Department of State Parks and Cultural Resources)

Wool sacks being freighted to the railroad at Moneta, Wyoming. A four-wheeled cooster wagon brings up the rear. (Photo courtesy Wyoming Department of State Parks and Cultural Resources)

together with a short tongue. The cooster, the last wagon in line, carried no freight—and for good reason: it was Harry's 'home on wheels.' Much the same as the small covered wagon used by a sheepherder, it contained everything Harry needed when he made camp for the night."[30]

George Coleman, a freighter in the Big Horn Basin area of central Wyoming, left a photographic record of his freighting days which often involved long trips to Casper. In the photo of Coleman's wagon shown on page 50, the door and the stovepipe appear to be backward. At first glance, one might assume the image has been reversed; however, the tent is stamped "Omaha Tent and Awning Company" which reads correctly from left to right. Why were the cooster's door and stove opposite those two items in a sheepwagon?

Suppositions might include that perhaps this model was built for a left-handed person since the brake inside the normal wagon was always on the right side, making it awkward for a southpaw to maneuver. According to some accounts, the reason

Mrs. George Coleman, a freighter's wife, in the door of a cooster wagon. Exterior features that distinguished the cooster wagon from the sheepwagon were the position of the door and the stovepipe which were the reverse of the sheepwagon. This cooster wagon also has a front window. (Photo courtesy Wyoming Department of State Parks and Cultural Resources)

A freight team with a cooster wagon hauls lumber to the newly discovered oil field at Salt Creek, Wyoming, circa 1915. (Photo courtesy Wyoming Department of State Parks and Cultural Resources, Coleman Collection)

for reversing the outfit was so a freighter would never be mistaken for a lowly sheepherder! One could tell the two apart at a distance by the position of the door and the stovepipe.[31]

Apparently, not many coosters were built since they are now quite rare and not seen on the range today. It is an interesting fact that although all sheep ranchers have of course heard of the sheepwagon, many have no knowledge of its close relative used by freighters, the cooster wagon. The rapid growth of transportation technology no doubt accounts for their relative obscurity. The rise in the sheep business coincided with the large-scale construction of railroads that soon criss-crossed the country, linking markets much more quickly than horse-drawn wagons ever could. Although freighting continued into the early twentieth century, the development of the automobile and truck rendered that once-essential lifeline to civilization obsolete.

CHAPTER FOUR

Blacksmiths, Builders, and Women's Work

BY 1900, AS THE SHEEP business continued to expand rapidly throughout the western states, sheepwagons were in demand. In the last days of 1899, a Douglas, Wyoming newspaper reported, "The modern sheep wagon is a familiar sight to most of our people—dotting our vast expanse of plain and valley, or passing to and fro through the city occasionally... Florence Hardware, of this city, has turned out 38 similar wagons during the year and yet are unable to catch up with their orders."[1]

A rancher had a choice as to where he could obtain a sheepwagon. He might build one himself; one contemporary Wyoming rancher recalled that his uncle made a number of the family's fifteen sheepwagons as late as the 1930s. This same man also supplied his design to a local lumber company that made sheepwagons.[2] But perhaps the most common way a rancher obtained a new sheepwagon was through the local blacksmith, and every town had at least one of these valuable craftsmen. Blacksmiths played a key role not only in the invention and modification of the

Thirteen sheepwagons, circa 1920s, lined up at Warren Livestock's Pole Creek Ranch north of Cheyenne, Wyoming. During this period, Warren Livestock ran approximately 27,000 sheep. (Photo courtesy Wyoming Department of State Parks and Cultural Resources)

H. H. SCHWOOB

WAGONS and FARM MACHINERY

REPAIRING and GENERAL JOBBING

Sheep Wagons, Wagons and Buggies Built to Order.

Horseshoeing a Specialty Satisfaction Guaranteed

CODY, WYOMING

G. McLAUGHLIN,

General Blacksmithing.

Wagon, Carriage and Plow Work.

Builder of Sheep Wagons : : : :

ODY - - - - - - WYOMING

Advertisements for sheepwagon builders in the Wyoming State Business Directory, 1907 (Source: Wyoming State Business Directory)

first sheepwagons, but also in the subsequent building and repair.

Blacksmiths who doubled as sheepwagon builders began to publicize their products in the early 1900s with advertisements in state business directories. These ads continued until around 1920. "J. C. Jacobsen, Builder of Sheep Wagons To Order" proclaimed one, while blacksmith F. L. Belcher's read "Manufacturer of the Belcher Sheep Wagon. Give Me a Trial." [3]

Newspapers also ran sheepwagon ads, such as the two that appeared for rival builders in a 1907 Wyoming newspaper, featuring photographs of their respective models. The Florence Hardware Company's ad noted "...Where We Make the Celebrated Florence Sheep Wagon." [4]

The blacksmith/sheepwagon builder might construct the entire unit himself, or perhaps work with a wagonwright or carpenter who built the wooden wheels and box. Or the running gears could be purchased separately from such companies as the Bain Wagon Company, the Studebaker Brothers Manufacturing Company, Winona Manufacturing, or the Peter Shetler Company. Mass-manufactured equipment like this arrived from those eastern companies

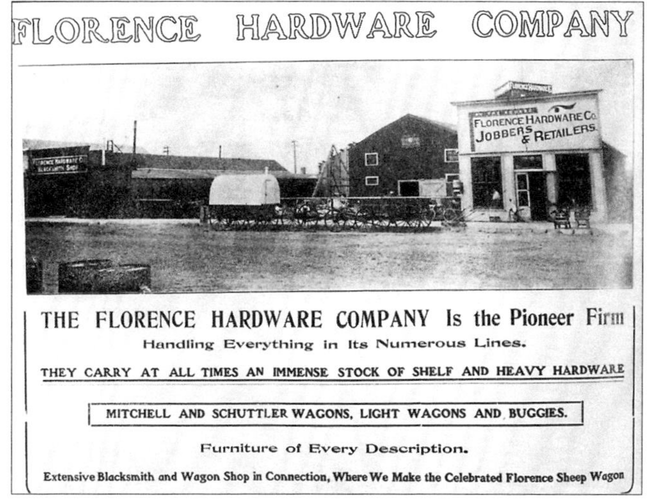

Advertisements in the Douglas, Wyoming newspaper, 1907. Two prominent sheepwagon builders worked in this small Wyoming town known as a regional sheep center. The A. & A.C. Rice wagon, called the "Douglas Sheepwagon Bed," featured a side canvas flap for greater ventilation. This was an unusual feature for sheepwagons. (Source: *Douglas Budget*)

by railroad. The local hardware store usually stocked small cast-iron stoves such as the Handy 714L or 816L, popular models often installed in sheepwagons, or those made by the Great Western Stove Company. Other sources of stoves were catalogs such as the popular Montgomery Ward's.

A very large sheep operation such as the Warren Livestock Company of Cheyenne, Wyoming, employed two English-born blacksmiths full-time during the 1920s and 30s. Their job included the building and repair of the ranch's forty-plus sheepwagons. Although Warren was one of the largest sheep ranches in the west, many ranchers had ten or fifteen working wagons, which translated into ten to fifteen herders caring for twenty to thirty thousand sheep.[5]

Women's Work

It is doubtful that any women actually built sheepwagons, as blacksmithing and carpentry were men's work, but they certainly contributed to the final product. At age ninety-one, Vashti Huff, daughter and wife of sheepwagon builders, shared her vivid memories of

Vashti Henderson Huff (second from right), was the daughter of sheepwagon builder Jimmy Henderson and the wife of sheepwagon builder Elmer Huff (second from left), Buffalo, Wyoming. Vashti helped Elmer build sheepwagons in the 1940s. (Photo courtesy Richard Francovic collection)

the building process. Vashti "was used to people living in wagons" long before she saw her first sheepwagon. Raised in Alabama, Vashti remembered that bands of gypsies, living in wagons, were part of the local landscape and some families returned every year to camp on the outskirts of her town.

Vashti's father, blacksmith Jimmy Henderson, arrived in the town of Buffalo, Wyoming in the early 1920s and went into the business of making sheepwagons. By this time, Buffalo had become the center of Basque sheep ranching in the state. Herders who had arrived in the town ten years earlier from the Basque country of the Pyrenees Mountains between France and Spain had become successful sheep ranchers with large flocks that summered in the Big Horn Mountains. The business was growing in the area, along with a demand for that essential tool of the trade, the sheepwagon.

Henderson was joined in his enterprise by Vashti's husband, Elmer Huff, in the early 1930s. Henderson eventually left the area but his son-in-law, a skilled and creative blacksmith, continued as sheepwagon builder par excellence in the area until his death in 1960. Many of the wagons Huff built for the local Basque ranchers, and converted to rubber

Jimmy Henderson outside his blacksmith shop in Buffalo, Wyoming, circa 1930. (Photo courtesy Richard Francovic collection)

Blacksmith Jimmy Henderson with the parents of the famous Marx Brothers. Mr. and Mrs. Marx were photographed with Henderson and one of his wagons. The Marx couple passed through Buffalo en route to Yellowstone National Park, circa 1928. (Photo courtesy Richard Francovic collection)

tires in the 1950s, can still be seen today parked on lawns and alleys in Buffalo, Wyoming.

No two sheepwagons were built exactly alike and Vashti remembered that although her father taught his techniques to her husband, Huff developed his own ideas as time went on. For example, Henderson insulated his wagons with only one blanket sandwiched between an inner and outer layer of canvas. Huff believed two blankets provided more protection from the elements; he also used the more practical oilcloth to line the interior roof. It was easier to clean than canvas and often patterned, which brightened a herder's home on wheels.

Vashti assisted her husband in the finishing of these covers. An accomplished seamstress, Vashti sewed blankets together to form the double layer of insulation between the outside canvas cover and an interior oilcloth. It required dexterity to join two blankets together without bunching them up in the sewing machine.

The challenge was to make the cover as weatherproof as possible for the harsh climate of the western states. The size of the early-day canvas meant that two pieces were needed to cover the bow top. Harder to sew than the blankets, the canvas wouldn't flow

Blacksmiths like Jimmy Henderson often made repairs on sheepwagons. (Photo courtesy Richard Francovic collection)

smoothly through Vashti's machine so Elmer farmed out that job to a local saddlemaker.

A seam in canvas, as in any material, creates a weak spot. Elmer's method of making a weatherproof cover involved overlapping the two pieces of canvas and creating two seams rather than one to keep the material from leaking where joined. This double seam was then carefully placed to lie on the roof somewhere between the bed and stove below. This practical measure helped further ensure that when the canvas did leak, as it eventually would, the two crucial areas of the interior—the stove and the bed—were not directly under the weak seam.

Vashti recalled a waterproofing technique of Elmer's, one he learned from his job on an Oklahoma road crew when he and Vashti lived in a tent. Elmer heated gasoline over a fire and when it reached a certain temperature, he added paraffin wax. The hot gasoline eventually evaporated and he mopped this mixture on the canvas top, already installed on the wagon, to provide a layer of waterproofing. When improvements to the canvas fabric made it more durable, Elmer discontinued this process.[6]

Similarly, Zella Pelloux Slayton remembers working with her mother to sew together a blanketed layer of insulation for her rancher father's sheepwagons. Zella recalled that unlike other wagons, however, they insulated only the area above the bed.[7]

Apparently, this practice of using blankets for insulation, single, double, or covering only one-half the wagon, was not universal, a fact which points up the regional variations. A contemporary sheepwagon expert, who grew up with the wagons in Montana in

Mrs. Martin Pelloux, circa 1911, wife of a Buffalo, Wyoming sheep rancher, sewed blankets together for insulation for her husband's sheepwagon. Mrs. Pelloux stands in the door of the wagon. This wagon has a distinctive awning over the front door. (Photo courtesy Johnson County Library, Buffalo, Wyoming)

the 1950s and 60s and restored his first one in the 1970s, scoffed at the idea of blanket insulation; it was never used on wagons with which he was familiar. He maintained how impractical this method would be: when the blankets got wet, during a storm, they would become hard to dry and also stink. The wagons he remembered had only two layers of canvas for insulation. It is indisputable, however, that some type of blanket insulation was common in Wyoming because so many people have made note of it.

According to Vashti, a local woman suggested the idea of a "honeymoon wagon," a larger, twelve-by-seven-foot wagon, to her husband. The wife of a Basque rancher came to Elmer's shop one day and asked him if he could build such a wagon. Huff replied that he could build whatever a customer wanted. This woman lived in a wagon with her husband on the Big Horns summer range. What she wanted, she told Elmer, was a wagon "big enough to put my hats" for which she was famous. Huff obliged and built one with extra room for her hats. During the 1950s, he built five or six more of these large wagons, referred to locally as honeymoon wagons.[8]

The honeymoon wagon was always a luxury item built for the rancher, not the herder. The additional

Two sheepwagons with canvas tops. The wagon on the right appears to have a second layer of canvas. Since this is a winter scene, it was most likely used as insulation. A ranch brand is drawn on the canvas on the wagon on the left. The height and width of stovepipes varied greatly on sheepwagons. (Photo courtesy Wyoming Department of State Parks and Cultural Resources)

foot or more of length and six inches of width made a remarkable difference in the interior space. It accommodated more people, often the family members, more comfortably, with the space between the door and the bed receiving most of the additional length. Of course, a wagon this large was heavier and more awkward to move. However, the rancher's wagon probably didn't move as often as the herder's smaller wagon.

It's likely that the honeymoon wagon, built on a rubber-tired chassis, would not have been practical before the 1950s when pickup trucks began to replace the horses that previously pulled the sheepwagons to the mountainous summer range. Before this time, sheepwagons were not built larger because the standard size was the most manageable in length, width, and weight for horses on rough terrain.

The Basque sheep ranchers in Huff's town of Buffalo, Wyoming still speak respectfully of the talented blacksmith who built so many of their sheepwagons. One such rancher paid a moving tribute to Elmer Huff in the obituary she wrote: "He was a true artist with tools...Even the prairies are dotted with memories of the deceased as most all of the sheepwagons in this area were built by Mr. Huff, or his predecessor and late father-in-law Jim Henderson."[9]

10

An Ahlander "Home on the Range" wagon with a side window, parked on the Laramie Plains. Two dogs rest under the wagon. (Photo courtesy Wyoming Department of State Parks and Cultural Resources)

CHAPTER FIVE

The Commercial Manufacture Of Sheepwagons

ALTHOUGH INDIVIDUAL builders made many sheepwagons, commercial manufacture of the wagons had begun by the turn of the twentieth century. A rancher could order a sheepwagon from a catalog, such as the one issued by Studebaker Brothers Manufacturing Company of South Bend, Indiana, a famous maker of all types of wagons. Beginning as a blacksmith shop in 1852, Studebaker rose to prominence among wagon companies in the late 1850s when it received its first big army contract to produce a hundred vehicles, both six-mule wagons and escort wagons, to transport supplies to the Plains. Other companies also manufactured these two models for the military; however, Studebaker built the majority of them after 1858. Studebaker claimed to be the largest wagon manufacturer in the world by 1885, the year they produced over 75,000 wagons.[1]

Studebaker manufactured sheepwagons from 1899 until 1913.[2] Their version was called the "sheep camp bed," a model that changed very little during its years of production. It measured eleven feet long

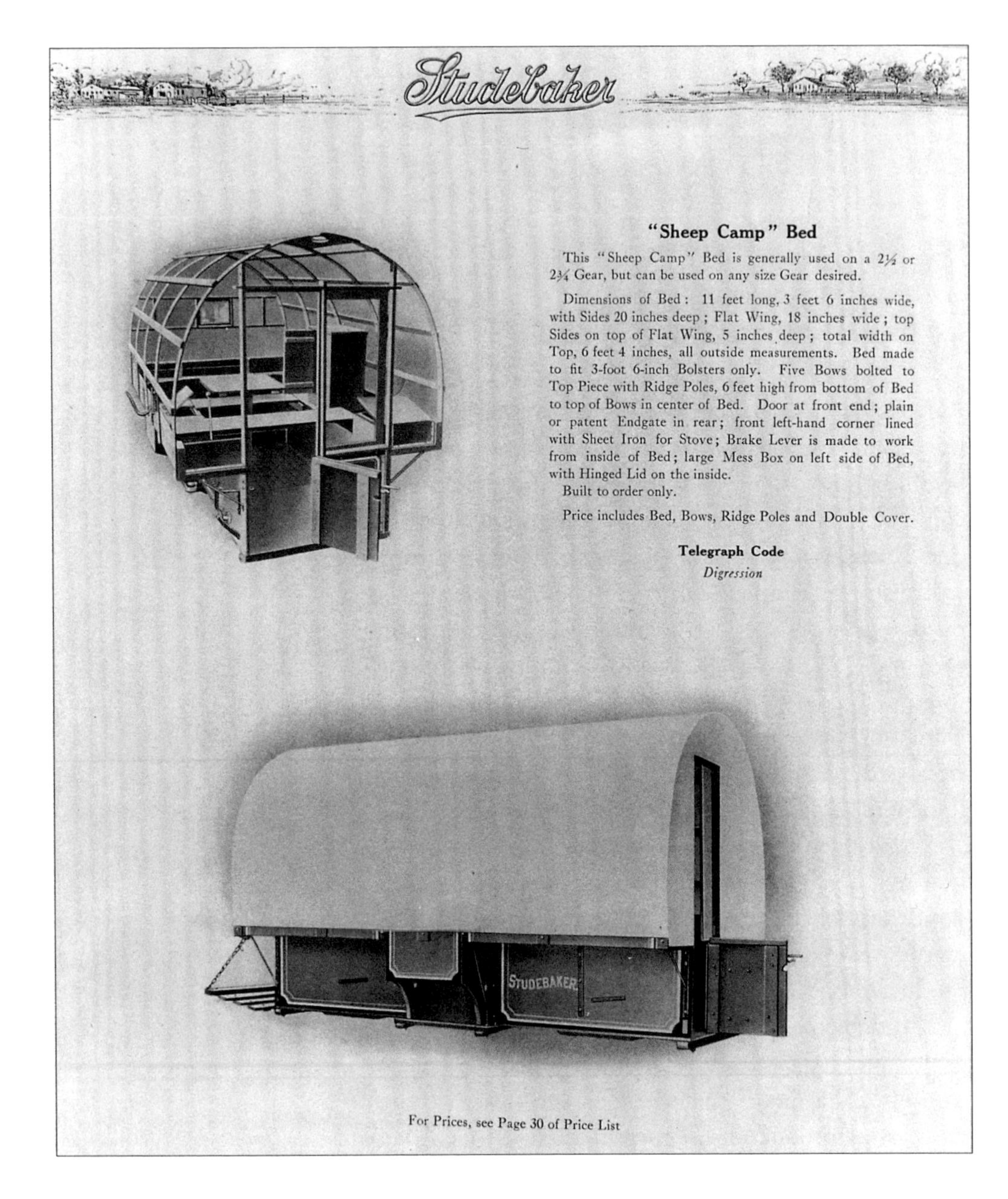

Studebaker

"Sheep Camp" Bed

This "Sheep Camp" Bed is generally used on a 2½ or 2¾ Gear, but can be used on any size Gear desired.

Dimensions of Bed: 11 feet long, 3 feet 6 inches wide, with Sides 20 inches deep; Flat Wing, 18 inches wide; top Sides on top of Flat Wing, 5 inches deep; total width on Top, 6 feet 4 inches, all outside measurements. Bed made to fit 3-foot 6-inch Bolsters only. Five Bows bolted to Top Piece with Ridge Poles, 6 feet high from bottom of Bed to top of Bows in center of Bed. Door at front end; plain or patent Endgate in rear; front left-hand corner lined with Sheet Iron for Stove; Brake Lever is made to work from inside of Bed; large Mess Box on left side of Bed, with Hinged Lid on the inside.

Built to order only.

Price includes Bed, Bows, Ridge Poles and Double Cover.

Telegraph Code

Digression

STUDEBAKER

For Prices, see Page 30 of Price List

A page from the 1912 Studebaker Brothers, Mfg. Co. catalog. Studebaker produced sheepwagons from 1899 until 1913. (Photo courtesy Studebaker National Museum)

and six feet three inches wide, had five bows, a three-and-a-half-foot wide bed, a rear window, a table that pulled out from under the bed, and a large mess box with a hinged lid on one side.

The Studebaker sheepwagon sold for $110 ($95 without the canvas cover) and included the bed, bows, and ridge poles. The running gear cost an additional $122. In 1902, a buyer paid extra for "bunk boards, table, canvas cover." By 1912, the base price included a "double canvas cover, additional mess boxes, racks, etc., according to specifications." For an extra eight dollars, a buyer could order a "brake lever…made to work from inside of the bed."

The company also sold an assortment of running gears for the sheepwagons in their catalogs. Advertisements stated "the 'sheep camp' bed is generally used on a 2-½ or 2-¾ Gear, but can be used on any size Gear desired." A buyer could choose among a variety of running gears such as the "General Teaming Gear," the "Block Tongue Double Reach Gear," or others.[3]

Although the basic wagon shell was standard, Studebaker advertised their camps as "built to order," personalized according to the wishes of the buyer. This customization by both the builder and the buyer was a hallmark of the versatile sheepwagon, a tradition that continues even today with the few modern wagons still made.

Studebaker used two paint colors on the majority of their wagons: green for the wagon box, red on the running gears.[4] This color combination became the most popular for all sheepwagons, although the A. & A. C. Rice Company of Douglas, Wyoming, painted at least some of their wagon boxes blue with yellow pinstriping on all flat surfaces, and vermillion running gears. Rice wagons bore the Rice name stenciled on the rear of the box, and its number on one of the outside grub boxes. It is unclear if the numbers corresponded to the total number of wagons built, but a number 234 Rice wagon has been seen which may indicate the company built at least that many. Studebaker also numbered its sheepwagons; number 692 belongs to a man in Idaho.[5] Ranchers often painted their brand on the outside of the wagon, usually on one of the grub boxes.

Studebaker had branch stores in western cities. An advertisement placed in the 1906–1907 Wyoming State Business Directory for one such store in Montpelier, Idaho noted "sheep wagons a specialty."[6]

The fact that the Studebaker catalog referred to the wagon as a "sheep camp" while their Idaho branch

Sheepwagon builders often bought canvas from companies like the Kistler Tent and Awning Company. This "ghost" sign advertising Kistler can still be seen on a building in downtown Casper, Wyoming. (Author's photo)

called it a "sheep wagon" is curious. No definitive source has been found that explains this use, although "camp" may be a holdover from the early days when the herder slept on the ground or in a tent. One can also speculate that the term "camp" was a Utah influence, possibly of Mormon origin. Throughout that state "camp" is still the common word and the latter-day manufacturers of the wagon all refer to it by that name. Both terms are used in Wyoming, depending on the region; in such Mormon areas as the Big Horn Basin and southwestern part of the state, "camp" is the preferred term. The wagons are also known as "arks" in an area of eastern Oregon.[7]

Studebaker wasn't the only company manufacturing sheepwagons in the early 1900s. Utah was home to a number of smaller companies that produced them. By 1910 and perhaps even earlier, the Sidney Stevens Implement Company of Ogden made a version they named the "Home Comfort" camping wagon which could be shipped in sections and assembled at its destination. It cost between $550 and $650 depending on the type of finish and equipment. This wagon's table did not slide from under the bed; instead, it could be latched in an upright position to the large cabinet built on one of

The famous "Home on the Range" sheepwagon was built by the Ahlander Manufacturing Company of Provo, Utah. Some ranchers referred to these as "bread loaf" wagons. Ahlander was the first company to make wagons specifically for a rubber-tired running gear. This wagon is located on a ranch near Opal, Wyoming. (Author's photo)

Right: An old "Home on the Range" wagon. Ahlander built approximately three thousand of these wagons between 1920 and 1976. The flat sides allowed for more interior storage room. (Photo courtesy John Cowan)

the side benches. An outside rack carried fuel oil for the interior lamps.[8]

Another such firm, the Ahlander Company of Provo, Utah, produced a modernized version of the sheepwagon beginning in the early 1920s, known as the "Home on the Range" Sheep Camp Trailer and usually built on a rubber-tired Model-T Ford frame. Sheet metal replaced the canvas top, and the wagon's wide rectangular shape, with flat bottom and sides, allowed for more storage space within since it did not have to accommodate the narrower, traditional wagon box and the high wooden-wheeled running gear. The wagon, which resembled a compact loaf of bread, rode lower to the ground than the traditional model and was often referred to as the "Mormon wagon" because of Utah's large Mormon population which included many in the sheep business. The Ahlander Company built approximately three thousand "Home on the Range" models between 1920 and 1976, when the firm ceased production of their camps.

Another Utah manufacturer, Wm. E. Madsen & Sons Sheep Camps, also began building in the 1920s and operated until 1970, when they were bought out by Eddy & Sons who continued to produce the

The most innovative feature of the Ahlander "Home on the Range" wagon was its rubber-tired running gear. This tire is original to the model (Author's photo)

Even sheepwagons change with the times. As ranchers made the transition from horses to trucks, a rubber-tired chassis replaced the original running gear which made the wagons easier to pull. This trend accelerated after World War II. (Photo courtesy Louise Turk collection)

"Madsen Home" until at least 1978. Madsen used the catchy phrase "When Better Camps Are Built, We'll Build Them" in their marketing material. The Madsen camp resembled the Ahlander version with sheet-metal siding and a rubber-tired chassis.

A third Utah company, Olson Brothers of Mt. Pleasant, operated until as late as 1980. The Olson camp featured two beds, insulated floor and walls, and interior wood paneling. Resembling the Ahlander and Madsen wagons, the Olson model was mounted on rubber tires and had aluminum siding. "Builders of Quality Camps," the company touted its wagons as "built to last 20 years," "easy to clean," and "pull as fast as you like." Like Ahlander and the Eddy/Madsen companies, Olson Brothers is no longer in business since the sharp decline in the sheep business around 1980.[9]

The Schulte Hardware Company of Casper, Wyoming also produced a type of sheepwagon made for rubber-tired running gear, similar to Ahlander's "Home on the Range" wagon, which became informally known as the "Casper wagon." This company stopped making sheepwagons around 1942 when the second-and third-generation family firm diversified into other businesses including a successful car dealership.

A number of sheepwagon builders lived in Utah. The Olson Brothers built a wagon similar to the "Home on the Range" model. The Olson wagon was called a "Sheep Camp." (Source: *National Wool Growers Magazine*, 1980)

Right: Wilson Camps, Inc. of Midway, Utah is the only manufacturer of sheepwagons in the United States today. Although the exterior resembles a modern camper, the interior configuration is very similar to the tradition sheepwagon. The term "sheep camp" is commonly used in Utah. (Source: Wilson Brothers, Inc. sales brochure)

Above: Wilson Camps, Inc. places this metal plate on all its wagons. (Author's photo)

The only firm in the United States that commercially manufactures sheepwagons today is Wilson Camps, Incorporated of Midway, Utah. Established in 1979, the company consists of octogenarian Emer Wilson and his two sons, Doyle and Mark. Wilson Camps produces rubber-tired, customized sheepwagons in three sizes: each is seven feet wide and can be built twelve, fourteen, or sixteen feet long. The wagon prices start at $9800 and go as high as $15,000 for the top-of-the-line model which includes such deluxe features as an additional roll-out bed, solar, electrical and propane packages, and water system. The company makes fifteen to twenty-five wagons a year.

Wilson Camps makes half of their wagons for non-ranching uses, selling them "to every walk of life," according to Doyle Wilson. The federal government bought some of the camps to house trappers at a remote Utah experimental station. A National Park Service employee lived in a Wilson camp while guarding Indian artifacts from looters in southern Utah. Artists and rockhounders buy them. Retired couples have purchased the camps for use as travel trailers. Lately, construction companies have been buying the wagons to park at work sites for lunch breaks and supply storage.[10]

PART THREE

THE HERDING LIFE

A Relic of the Past[1]

Behind the corral in sort of a heap
  Is a home like I used when I once herded sheep;
The roof is all gone, the sides in decay,
  The floor and the wheels are just rotting away.

The old doubletrees the horses hitched to
  Have long gone the way that good antiques do;
The tongue is a wreck, been all shortened up,
  Appears to have been stubbed for a modern pickup.

The bunk in the back where the herder slept
  And the bin underneath where the spuds were kept
Are weathered in place, like the table leg,
  And the outside shelf for the water keg.

The shelf over the bunk is still intact,
  But the glass in the window is loosened and cracked;
The hinged down door to the cupboard there
  Served as a table, a bench was the chair.

In the near corner next to the old door
  Is that sheepherder stove all rusted with lore;
Baked the best biscuits, cooked the tastiest stew,
  Boiled better coffee than Mother could brew.

The overhead bows are good to this day,
  But the canvas cover has wasted away;
The lantern for light is still hanging there,
  With a candle nearby as a needed spare.

Hasp on the door is latched with a stick,
  Strangers could enter with nary a trick,
Could stir up a meal to their own desire:
  "Just clean up their mess, leave wood for a fire."

The last herder's camp I saw out on the range
  Was shiny, it glistened, it looked mighty strange;
With gas and refrig, who could dream of the like,
  And beside it was parked a Suzuki bike.

—James A. Ross
from his book *Saddle Up And Ride*

ATWATER KENT
RADIO
ATWATER KENT MANUFACTURING COMPANY
Philadelphia
Popular
Science
MONTHLY

CHAPTER SIX

Traditional Sheepherders

THE IMAGE OF THE sheepherder is every bit as defined as that of the cowboy, but his notoriety is of a different type. The herder never gained the widespread adulation and respect bestowed on the mythical broncobuster. The cowboy is an American original, the quintessential westerner. He's seen as a rough-and-tough individual, a laconic man of action, a one-hundred-percent male very attractive to women, a strong, fearless stoic most at home on a saddle miles from civilization. The cowboy was easily recognized, and later imitated worldwide, by distinctive clothing and such important accessories as boots, spurs, hat, bandanna, and oftentimes a fancy saddle.

Some contemporary writers have added a healthy dose of reality to the image of the cowboy in their more historically accurate portrayals of the cowpuncher's life on the range. In an essay that explored the cowboy as a Wyoming symbol, historian Robert A. Murray discerned three sub-species of the type: the "true old-time open-range cowboy of about 1867–1906," the ranch hand, and the movie cowboy.[2]

Herders read magazines and listened to the radio to pass the time and stay in touch with the outside world. A shotgun often hung near the bed. (Photo courtesy Buffalo Bill Historical Center, Cody, WY, Charles Belden Collection)

A typical Wyoming sheepherder, circa 1901. (Photo courtesy American Heritage Center, University of Wyoming)

Unlike most sheepherders, the open-range cowboy was only seasonally employed, working spring and fall roundups. In between times, he rode what became known as "the grub line," traveling from ranch to ranch doing whatever odd jobs he could in return for a hot meal and a roof over his head for a few days. According to Murray, he was "essentially a drifting...migrant laborer...uniformly underpaid, generally exploited." He often received less pay than the herder.

Murray described the second sub-species, the ranch hand, as a "transitional figure" between the open-range system, which he dated as ending in the early 1900s with the advent of widespread fencing, and modern ranching practices. Still with us today, the ranch hand "has been the basic laborer for settled and managed ranching...Ranch hands are far more stable personalities than the old-time range cowboy. They raise families, settle in a given community and seldom travel far away from it. And they just 'grab hold' and do whatever is needed to keep the place running. Some even become stockmen in their own right."

The third type, today's "drugstore cowboy," Murray described as popular by the early 1920s, "a wholly manufactured product of the movie industry sound stages. He represented a fusion and crystallization of ephemeral literary themes from the pens of Owen Wister, Ned Buntline, Bill Cody and other writers and showmen of the previous generation." Fueled by the burgeoning dude-ranch industry of the same period, and reinforced by television, the cowboy image resolved into the romantic western hero that many still believe populates the West. Today, the image has been exported abroad. In Germany, for instance, imitation western false-fronted towns and mock gunfights attract thousands of tourists eager to experience what is portrayed as "the Old West" with the mythical American cowboy as the main attraction and hero.

Like the cowboy, the sheepherder also suffers from a one-dimensional misrepresentation, although his reputation veers off in quite another direction. The western herder has usually been a figure of derision; the words "outcast," "crazy," and "drunk" were often applied to him and no article was ever written without the adjective "lonely" featured prominently in the text or title: "Sheepherder: The West's Lonely Hero,"[3] "Some Men Need It Lonely,"[4] and "The lonely life of a sheepherder"[5] are just a few examples. In his 1948 book *America's Sheep Trails*, still the most comprehensive history of the western sheep business, Edward N. Wentworth stated, "More misinformation

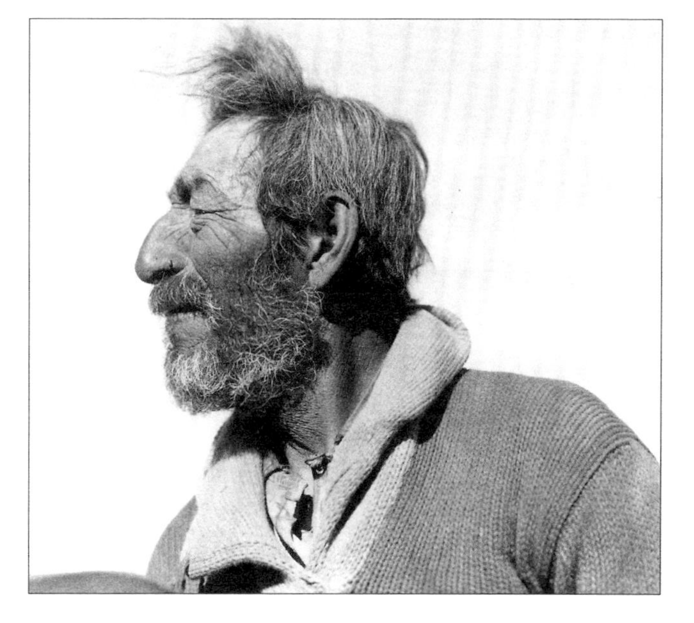

The weather-worn face of a sheepherder. (Photo courtesy American Heritage Center, University of Wyoming)

was current about him than any other class on the frontier, and the wildest vilifications of his character and personality were accepted as true."[6]

In the opening lines of his popular 1929 book *Sheep*, former herder Archie Gilfillan explained that "there are two general theories about herding. Some hold that no man can herd for six months straight without going crazy, while others maintain that a man must have been mentally unbalanced for at least

Archie Gilfillan, a South Dakota sheepherder and the author of Sheep. (Photo courtesy Middle Border Museum, Mitchell, SD)

six months before he is in fit condition to entertain the thought of herding."[7]

Although much of his book is written in a tongue-in-cheek style, Gilfillan's humorous take on herders reflected what many believed to be true about the herder personality. A man would have to be "mentally unbalanced" to choose and withstand the herding life. Herding was a low-man-on-the-totem-pole job in the West, in sharp contrast to the exalted biblical metaphor which likened Jesus to a shepherd, the spiritual guardian of his flock. Gilfillan noted that:

> *...time has dealt hardest of all with the herder. From a high and honorable place in the pastoral ages he has gradually descended, until in this age of industrialism he has only one consolation left, and that is the secure knowledge that he is working on rock bottom, that no matter what he does he never can get any lower, and that any time he makes a change in his occupation he will automatically rise in the social scale. You could fire a shotgun into the average crowd in the range country without hitting a man who had at some time herded sheep, but it*

would probably take the charge in the other barrel to make him admit it. About the only person who isn't ashamed to admit having herded is a sheepman, and he refers to it merely to show how far he has come.[8]

Various authors have explained the differences between the cowboy and the sheepherder by comparing cattle and sheep, and by analyzing human personality traits, region of origin, age difference, and ethnicity. So important are these perceptions that writer Bill O'Neal claimed they were a "primary factor" in the well-known sheep/cattle range wars which spanned a fifty-year period from the 1870s to about 1920. Wyoming has an especially notorious reputation because more sheep, sixteen thousand, were killed there than in any other state.[9]

The debate over whether cows are smart and sheep are stupid, or vice versa, has continued for over a hundred years in the West. They are two very different animals. Cattle are large beasts capable of surviving with little human oversight except for periods of roundup, winter feeding (which only began after the devastating 1886 winter), and calving. Sheep, on the other hand, are a smaller, more timid animal,

A sheepherder and his dog out on the range near Douglas, Wyoming, circa 1911. (Photo courtesy American Heritage Center, University of Wyoming)

Rancher Martin Pelloux (center) was born in France. By 1900 when this photo was taken, he was a Wyoming sheep rancher. The herder at right is a native of Italy. A variety of nationalities were represented in the early days of the western sheep business. Pelloux's brand is painted on the canvas. (Photo courtesy Zella Pelloux Slayton collection)

with a strong herding instinct, in need of human protection from a variety of predators and severe weather, especially during snowstorms when they tend to bunch up and smother each other. Perhaps because of their vulnerability, the herder became known for having a more maternal, caring way with his charges than the cowboy, although many a sick calf was saved by the care of the man in the big hat.

The cowboy was often depicted performing his athletic job in view of others, high atop a horse, his spurs flashing, chaps flowing, hat angled enough to shade his brow but still reveal a strong, resolute profile. The herder, on the other hand, was the solitary footsoldier, clothed in layers of dirty, bulky garb, plodding behind his flock. The perception was that the cowboy was young and vigorous while the herder was much older and sedentary.

In fact, many cowboys of the open-range era were young men who headed west in search of opportunities not available at home following the devastation of the Civil War. Among them were a good share of southerners, which may partly explain the origin of the cowboy's much-vaunted code of honor and chivalry. In contrast, the herder was oftentimes foreign-born, able to speak only broken English.[10]

This native-versus-foreign issue no doubt contributed to the enmity between the two groups. According to Edward Wentworth, "The majority of herders have been of so-called 'foreign' races. Mexicans and Indians predominated in the Southwest, while in California there were also Bearnais and Dauphinois French, Basques, Portuguese, and occasional Germans. During the period 1870–1890 there were many Chinese herders. In the Pacific Northwest, Scotch, Irish, and English herders appeared, as well as in Montana and Wyoming. Among the Mormons of Utah and southern Idaho, most of the herding was done by young men from the families owning the flocks."[11]

Racism also played a part in the elevation of one group and the denigration of the other. In a country not particularly known for tolerance of foreigners, the immigrant to the United States was often considered an inferior being: dirty, poor, and unable to speak the language. Sheepherding presented an opportunity for an ambitious newcomer to get a stake in the wide-open West, and was the first job for many who went on to become successful sheep ranchers.

In Wyoming, the first-generation Irish sheepmen of Natrona County imported herders from their native country. Irish-born men such as Patrick Sullivan and Mickey Burke established their sheep ranches in central Wyoming, then sent for herders from the old country. The Ellises, present-day sheep ranchers in the Casper area, got their start in Wyoming by working for an uncle, sheepman Tim Mahoney, who paid for their passage from County Cork, Ireland. These men, in turn, became ranchers and continued the tradition of recruiting herders from their native land.[12]

Immigrant Scotsmen also joined the sheep business. Robert Taylor, native of Scotland and one of the most prominent of the nineteenth-century western sheepmen, provided jobs for his fellow countrymen, who began as herders, progressed to partners, and eventually established their own successful sheep outfits. Taylor provided the start-up for a number of notable Scottish sheep ranchers in Wyoming, including his brother-in-law, W. T. Hogg, Robert Grieve, his nephew, George Taylor, and brothers Richard and David Young.[13] Noted western author Ivan Doig also wrote about first-generation Scottish sheepmen in northern Montana in his book *Dancing at the Rascal Fair.* In *This House of Sky: Landscapes of a Western Mind*, Doig related stories of the Finnish, Romanian,

Many Hispanic men from New Mexico and southern Colorado migrated north to work as herders in the western sheep business. (Photo courtesy American Heritage Center, University of Wyoming)

and Norwegian herders with whom he came in contact during his 1940s Montana childhood.[14]

A different tradition took hold in states along the Union Pacific Railroad line where ranchers employed Mexican-American herders (variously known as "Mexicans" or "Spanish people") from New Mexico or southern Colorado. Herding for a particular sheep rancher often became a family tradition, as a man introduced his son or brother into the business. A number of these Mexican-American families eventually settled in the towns along the Union Pacific in the early twentieth century. Today, people of Mexican origin in these towns can recall their fathers' or grandfathers' stories about their sheepherding days.[15]

A tough, solitary, and low-prestige job, herding attracted a variety of personalities for different reasons. For many men, the reason was simple: herding provided a job, a roof over one's head, and guaranteed food. Often little or no experience was required to become a sheepherder—only a willingness to give it a try. One learned the job best by doing it. Should the job become unbearable, a man could quit, although actually taking leave of the job presented some obstacles as the herder might have to wait for a visit from the camptender or rancher to get transportation to a

HERDER WITH DOG AND WAGON

This illustration was used in a Wyoming fourth-grade textbook written in the 1950s. (Source: *Wyoming's People*, Clarice Whittenburg, illustrated by Anne C. Mears, The Old West Publishing Company, Denver, CO, 1958)

town. Some stories do, however, tell of herders so miserable and desperate on the range that they just abandoned their woolly charges and walked to the nearest point of civilization.

The West's best-known sheepherder, author Archie Gilfillan, enjoyed the independence provided by a herding job. Like many before and after him, Gilfillan planned to be a herder for only a year "to think things over" before moving on to a new career different from those at which he'd failed. He ended up herding sheep for sixteen years in western South Dakota. A Phi Beta Kappa graduate of the University of Pennsylvania, Gilfillan used his time in a sheepwagon to read hundreds of books—Greek, Roman, and English literature—and write over nine thousand pages in his journal. His book *Sheep* is a thoughtful and witty look at the life of a high plains herder during the 1920s.

Loneliness is a nonstop theme in second-hand tales of a herder's life. However, a 1938 article in *The Southwestern Sheep and Goat Raiser* pointed out: "This loneliness of a sheepherder's life is often the theme of those who write without actually experiencing the life themselves, but sheepherders who have written of their own experiences...fail to mention it." [16]

Laura Bell, a sheepherder in Wyoming's Big Horn Mountains during the 1980s, expressed a similar sentiment: "People automatically think you're lonely. But you have your horse and dog for company, and the sheep. They put you through your paces emotionally. Why would I be out here if I wanted to talk to people all the time? I found myself getting real choosy about company. I really didn't want to see just anyone. I don't have that need." [17]

Of course, a herder's adjustment to solitude depended on his or her personality and interests. Ambivalent feelings about this solo lifestyle were perhaps the best one could hope for; an Oregon herder summed it up like this: "Oh, shucks I have been up to forty days without never seeing nobody. If I don't talk to anybody the rest of my life, I don't care. But I'm a reading hound. If I didn't have something to read, I'd go crazy. I have no use for a radio, I don't need no music. I'm the music myself...As far as company, why I look everywheres and I always have a full conversation going on. I talk to my dogs and horse and sheep. Don't think for a minute I don't want to talk to someone, but I have no one." [18]

A typical herder cared for a band of between fifteen hundred and three thousand sheep, a large investment for the rancher. Before 1900, that value in terms of breeding and wool represented as much as twenty-five thousand dollars per band; thirty thousand dollars in some cases by 1919–1920; and after World War II, forty thousand dollars.[19] The author of a 1956 book pointed out, "As the slightest irresponsibility might result in a serious loss, a high degree of loyalty to the job was necessary. Most herders, regardless of race or background, accepted their responsibilities, thus providing the profession with its share of unsung heroes who...have never been immortalized in song. Although frightened sheep never presented a danger comparable to a herd of stampeding cattle, many a herder has been frozen to death trying to save his flock in a blizzard."[20]

Clearly, the success or failure of a sheep operation rested heavily on the shoulders of these often maligned men. A good herder was prized by a rancher; if he stayed long enough he might become part of the extended ranch family and spend his retirement days in a small house or spare sheepwagon on the ranch, doing odd jobs in return for his room and board. Most sheep ranchers today have at least one story about the loyalty of such an employee, perhaps handed down through family lore or remembered from their childhood.

According to some, herding was never a boring job; others claimed just the opposite. Archie Gilfillan wrote:

> *One of the popular misconceptions about herding is that it is a monotonous job; or as a friend of mine put it, "Herding is all right if you don't have an active mind." But there is really little monotony in it. The sheep rarely act the same two days in succession. If they run one day, they are apt to be quiet the next. They herd differently in a high wind from what they do in a gentle breeze. They travel with a cold wind and against a warm one. They are apt to graze contentedly where feed is plenty and to string out and run where the picking is poor. Herding at one season is so different from herding at another as almost to constitute a different job. No one herding day is exactly like any other day, and there is doubtless much more variety in them than there is in the days spent in office or factory.*[21]

Another herder voiced just the opposite sentiment: "My job isn't hard. It's just the same thing day after day after day. It's the long hours and constant responsibility. In the summer you work 15-hour days. You get one hot meal a day if you're lucky, sometimes not even that. The only time the job is hard is when the weather is bad."[22]

While the job of herding may not have provided the brand-new day 365 times a year that Gilfillan described, it did offer variety during the four seasons. Spring was and still is a busy time in the sheep business, with the lambing season following closely behind the annual shearing (in the old days, this process was often reversed; shearing occurred after lambing). Although professional shearers were employed, the herder was on hand to help during this labor-intensive period. A shearing site, whether in a large shed by a railroad siding, at the home ranch, or elsewhere, might include all of the ranch's sheepwagons, drawn in from the winter range. After a long isolated winter, spring shearing season was anticipated not only by the herder but also neighbors and nearby townsfolk, as a time of activity and social interaction.

Herders were also very active during lambing time and helped consolidate the ewes and their young into bunches and then into bands that each herder eventually trailed to the summer range. The summer season in the mountains presented a new set of challenges. Storms occurred frequently and accounts of lightning striking a sheepwagon or killing a herder were not uncommon. The forested high elevations—sometimes so steep a herder packed in, leaving the sheepwagon behind until fall—required a different type of watchfulness than winter's wide-open range, and the herder might spend more time on horseback.

Sheep were usually trailed off the mountains sometime in September, preferably before autumn's first big snowstorm. Back at the lower elevations, the herder assisted with sorting out the sheep going to market and the formation of new bands for the winter. In the fall many herders took time off to spend a week or two in town. Rams were placed with the herd for breeding in the fall, then the herder trailed his band to the winter range and resumed his life in the sheepwagon, the yearly cycle complete and spring a long way away.

The herder as a drunken loner is one stereotype. It is true that some were men seeking escape from their past, their only hobby excessive drinking at every opportunity. One third-generation Wyoming sheep rancher remembered that his family never supplied vanilla extract to their herders because they invariably made wine from it or just drank it straight. He also recalled that many of the herders were Utah Mormons who did

Sheepherders and the cook (in wagon doorway) during the lambing season. This photo was taken shortly after a spring blizzard, not uncommon during lambing season. (Photo courtesy Buffalo Bill Historical Center, Cody, WY, Charles Belden Collection)

indeed have a drinking problem and had been ostracized by their families and the strict Mormon society.

This rancher also spoke of the "social network" in his town that linked ranchers, doctors, bartenders, and flop house owners during the 1940s and 1950s. If the herder was on a spree in town, the people in the informal network notified the rancher, who took charge of drying the man out and putting him back to work on the range a few weeks later.[23]

In his book *Sheepherder: Men Alone*, a fascinating study of contemporary herders, author Michael Mathes interviewed several herders about their town drinking sprees. One said, "Not all herders go on tears but most of them do. I asked Ed why. He said, 'You're craving excitement, something different. You're out there for months at a time without seeing anyone. You come to town and you don't know anybody. Where else are you going to go except a bar where you can meet some people?'"[24]

Elsewhere in the book, herder Ed reflected: "I'd say drinking's the biggest battle the sheepherders fought. Now why is somethin' I can't explain really. Your 'motions gets abuilt up and you jus' wanna lose all responsibilities and the only way to release 'em is a good tear—git good and sick. Now a good drunk lasts a coupla weeks, maybe a month."[25]

Sheepherders were a favorite subject for photographer Charles Belden. During the 1930s, Belden took a series of photographs of sheepherders relaxing in their wagons at his Pitchfork Ranch near Meeteetse, Wyoming. These posed photographs show the herders in typical leisure activities such as reading the newspaper, playing a guitar and listening to the radio. (Photo courtesy Buffalo Bill Historical Center, Cody, WY, Charles Belden Collection)

Another herder confessed, "My biggest bender was this last trip. Boy I'm tellin' you. I spent all told two thousand five hundred dollars. I mean I throwed it away. It only took me about fifteen days and I was busted...I go on a bender about every six, seven months."[26]

Mathes offered a thoughtful analysis for many a herder's excessive drinking. He speculated that their closeness to and caring ways with the sheep, dogs, and horses could not be shared in the company of other men because culturally males "had no history of warmth and nurturing. In fact male society threatened, with its very lack of emotion, all that herders had known of that side of their beings that merged with the country, its peacefulness, its depth of caring for itself and the things it held."[27] Although not everyone would agree with Mathes, his theory, based on direct observation, does offer an insight that goes beyond the simple assumption that most herders were misfits, alcoholics, and crazy.

For men confined to life in a sheepwagon on a remote range, any type of female sexual companionship, if only for a night, was elusive. Herders spent time with prostitutes in town when on their yearly spree. At other times, a rancher brought "girls" from town out to the herders when they were gathered at shearing camp. During the 1920s and 1930s, two settlements tucked away in the Big Horn Mountains, too small to be termed towns, catered to freighters and herders during the summer months, with liquor and prostitutes on the menu. Neither place exists today; a group of ranchers, irate over their herders frequenting the houses of ill repute, burned down one of these establishments.[28]

Another story is told of a man who set up a seasonal business using sheepwagons as a type of brothel for sheepherders when they were in the area. This man lined up twelve or thirteen wagons, with a girl in each one, along a country road in western Wyoming located near a popular sheep range.[29]

CHAPTER SEVEN

Sheepherding Practices and Tools

Using black sheep, a herder could keep track of the approximate number of animals in his care. Often bands contained one black sheep for every hundred white sheep. (Photo courtesy Buffalo Bill Historic Center, Cody, WY, Charles Belden Collection)

SHEEP PRODUCTION IN the western United States reached an all-time high around 1910. Individual state statistics for stock sheep that year give an idea of just how much the business had grown in four decades.[1]

Wyoming	5,480,000
Montana	5,385,000
Utah	2,742,000
Oregon	2,717,000

Sheep raising had become a very profitable business, and ranchers used a wide variety of methods, evolving since the early 1870s, in handling sheep, herders, and wagons.

For example, some ranchers believed in "lambing out" ewes before shearing them; others did just the opposite. Shearing often took place in a haphazard fashion, outside on the range where dirt mixed with the wool, producing an inferior fleece. Widespread reforms in shearing practices, modeled after Australian methods, were promoted shortly before 1920 in an attempt to make American wool more competitive with the better, cleaner grades of foreign wool.

Surprisingly, not all sheepmen approved of the sheepwagon, believing its use led to poor sheep handling. Because the wagons usually stayed in one place for a week or ten days, that meant the sheep had to return to the bedground over grass they had already grazed. One story tells of a Nevada rancher who even as late as 1933 hired only "tipi-tent burro herders because they moved their camp site daily while following the sheep."[2]

At this same time, progressive ideas were being applied to large industries throughout the country. Companies came under the scrutiny of the first generation of "efficiency experts" who analyzed the processes of production in order to improve upon them, believing standardizing tasks would minimize losses and increase prosperity. Reform ideas began to be proposed for the sheep industry as well.

Moroni Smith's Guide to Herding

Moroni Smith, a man with twenty-five years' experience in the sheep business, authored a book in 1918 titled *Herding And Handling Sheep On The Open Range In U.S.A.* in which he attempted to codify the western sheep industry. Smith opened his book by stating:

> *It has long impressed me that there has been need for some time to put into print the established facts so far as they have been proven to be reliable to the herding and handling of sheep on the open range...As we find there is a great deal of variation in the ideas of different herders, while they may be somewhat general on the common points, they vary a great deal in details, in fact so much that the results obtained from the labor of the most successful herders warrant consideration and with this idea in mind, it is to encourage the adoption of the most progressive methods in every move regarding the handling of sheep.*[3]

Smith packed his slim, fifty-one-page volume with all manner of practical advice for the rancher and herder from "Getting the Right Start In the Morning" and "The Proper Way to Turn Sheep" to a month-by-month almanac of range conditions regarding water, feed, weather, and predators. For example, he advised that in December, when supplemental feeds might be necessary, band sizes should not exceed fifteen hundred sheep because a good herder could not properly feed more than that.

Smith devoted a separate section of his book to "Care of Horses and Camp Equipment" which included tips for sheepwagon handling. He advised, "All sheep wagons must have a general going over; tighten all the bolts on gears and clamps; see that the neck-yolk and double-trees are serviceable and strong at all times; keep stay-chains on every wagon."[4]

He stressed the importance of good horse care in order to make the job easier for those animals when moving the wagon.

> *An old expert teamster's advice to keep a good pulling team true was to always set the wagon with a little downhill to start, never solid start; no matter how true your horses, never ask them for a heavy pull the first pull of the day, and especially in the morning. Locate a place for the wheels to set so as to put the camp with a little lower ground to the front end and all the wheels about level. This is not only an advantage to the team, but is convenient to the conditions of camp work...In pulling up heavy grades or steep hills, the horses must be rested often enough to keep them from getting out of wind.*[5]

Careless handling of equipment was inefficient and wasteful. Smith counseled: "There are some men that get into the habit of leaving a part of the camp equipment or some of the grub supplies on old camp grounds, planning to return later for it, usually neglect to ever get the cached articles. There is no excuse that will justify a policy of this kind and this practice cannot be allowed. Should any unnecessary equipment or grub accumulate around the camp it should be arranged to be taken to headquarters."[6]

Since control of expenses for sheep camps would naturally be an important factor for any sheep operation, Smith cautioned, "All equipment must be purchased by the manager or duly authorized person to insure getting the proper kind and type of equipment. No purchase should be made by any other person until they have obtained authority to purchase certain articles. Exceptions may be made in case of emergency."[7]

Smith's sentiments on this matter are echoed in a letter written in 1925 by prominent Wyoming sheepman Thomas Seddon Taliaferro to his twenty-seven-year-old son who was running one of the family's sheep ranches: "With the organization that you

Smith's guide also provided a list of recommended equipment for sheep camp. (Photo courtesy American Heritage Center, University of Wyoming)

REGULAR EQUIPMENT FOR SHEEP CAMPS

6 Knives and Forks	1 Shovel	Set Harness
6 Plates	1 Pick	Sheephook
6 Teaspoons	Pack Saddle	Horse Blankets
2 Tablespoons	Ropes	Tub for water
4 Cups	Blanket-Pack Bags	Kegs, or milk cans
1 Bread Pan	Riding Saddle	Sack Needle Twine
1 Small Enamel Pan	Hobbles	Nose Bags
2 Saucepans	Horse Bells	Lamp and chimney
2 Dippers	Seamless Sack	Dish Cloths
1 Wash Dish	Lantern, Coal Oil	Stove and Pipes
1 Tent	Gun, ammunition	1 Axe
1 Pr. Sheep Shears	Chains	½ Gal. Granite Jar
1 Meat Saw	Branding Irons	Paint and Oils for
Sheep Salt	Oats for horses	branding.
Bedding	Powdered Rosin	
Tent	Sheep Wagon	

One quart can Standard make Coal Tar Sheep Dips.

One quart Raw Linseed Oil to reduce the coal tar dip with half and half for application to sore mouth on sheep, saddle sores, general disinfectant and antiseptic.

have, the promiscuous buying of groceries by your camp movers must stop immediately...There is no earthly reason why your Foreman cannot notify you once each month of what he wants, and get you to order it...Your grocery bills show a recklessness in purchase that must be stopped...It ought not to be necessary for me to explain, or to reason, or to argue this proposition."[8]

Food

Moroni Smith also included detailed lists of equipment and groceries for sheep camps. His suggestion of such essentials as six each of knives, forks, plates, and spoons indicated that the herder should expect to feed more than just himself and the camptender on some occasions. The list of groceries shows just how deficient the herder's diet was in fresh fruit and vegetables.[9]

Naturally, food, both the cooking and storing of it, would be a major concern in the deserted range, many miles from a town. Assured a steady delivery of food every seven to ten days by the camptender, the herder was on his own as to efficient storage and use of the somewhat bland food. To keep the limited amount of fresh food from freezing, sheepherders

Moroni Smith's 1918 guide to herding provided a typical grocery list. (Photo courtesy American Heritage Center, University of Wyoming)

ON THE OPEN RANGE 57

REGULAR GROCERIES AND SUPPLIES FOR CAMP.

Flour
Sugar
Tea
Pepper
Soda
Tooth Picks
Beans
Milk
Oat Meal
Wheat Flake
Nutmeg
Pickles
Yeast Cake
Butter
Rice
Dried Fruit:
- **Apples**
- **Peaches**
- **Prunes**

Coffee
Table Salt
Baking Powder
Matches
Onions
Sego
Canned Goods:
- **Tomatoes**
- **Peas**

Lemon Extract
Honey or Syrup
Coal Oil and Candles
Potatoes
Soap:
- **Toilet, Laundry**

Towels

Sheep pelts or other miscellaneous property must not be sold or disposed of, only by the manager or duly authorized person. All equipment must be purchased by the manager or duly authorized person to insure getting the proper kind and type of equipment. No purchase should be made by any other person until they have obtained authority to purchase certain articles. Exceptions may be made in case of emergency.

All old equipment must be returned to headquarters for disposal.

"rolled up in their blankets a sack of fresh root vegetables such as potatoes, onions, carrots, and cabbage, but they seemed not to mind the dirt which sifted through their bedding from the vegetables."[10]

One woman who lived in a sheepwagon with her husband recounted that he, "In common with most range men...was skilled in cooking a few simple plain foods which men eat in camp...He managed to provide us with a better variety of foods than the usual sheep camp fare by digging a small cellar in the bank of a gulch nearby, over the opening of which he hung some horse blankets, and inside placed a lighted kerosene lantern to keep from freezing some fresh and canned foods."[11]

Tools of the Trade

Besides the sheepwagon, a herder also needed a variety of equipment and animals that helped him perform his job.

COMMISSARY OR SUPPLY WAGON. A supply wagon, also known as a commissary (and sometimes called a hooligan), was an important piece of equipment in the sheep business. Of simple utilitarian design, about ten feet long with flat sides and open on top,

HONEST SHEEPWAGON CARROT CAKE

Makes 1 or 2 cakes

Remember to start this cake the day before you want to serve it.

- **1⅓ cups granulated sugar**
- **1⅓ cups water**
- **1 cup raisins (or chopped candied fruit)**
- **1 tablespoon butter**
- **2 large carrots, finely grated**
- **1 teaspoon ground cinnamon**
- **1 teaspoon ground cloves**
- **1 teaspoon ground nutmeg**
- **1 cup chopped walnuts**
- **2½ cups sifted all-purpose flour**
- **2 teaspoons baking powder**
- **½ teaspoon salt**
- **1 teaspoon baking soda**

In a medium saucepan put sugar, water, raisins, butter, grated carrots, cinnamon, cloves and nutmeg. Simmer mixture for 5 minutes, then cover and let it rest for 12 hours. Why it gets so tired is one of those little mysteries. But do it.

Then add walnuts, flour, baking powder, salt and baking soda. Mix it all up. Bake it in 2 oiled 8-by-4-inch loaf pans or 1 tube pan at 275 degrees for 2 hours. Cool, then wrap it in foil.

A good-tasting, rich-looking, moist, sturdy pioneer cake this is, and good for every meal including breakfast.

Recipe for honest sheepwagon carrot cake. (Source: *The Oregonian*)

the supply wagon could often be seen on the trail hitched behind the sheepwagon. The wagon carried the essential water barrels, hay, and other supplies needed on the range far from town. (In the early days, the herder's only water supply was a nearby spring or stream.) Once the sheepwagon was set in place the supply wagon was unhitched and might stay at the campsite or perhaps the camptender would take it back to town. The supply wagon was usually painted green or sometimes yellow.

A commissary wagon can be seen in the photograph on pages 146-147.

SHEEP HOOK OR SHEPHERD'S CROOK. The sheep hook, as it is called in the western United States, is a wooden staff about six feet long with a large curved metal hook at one end. One often sees the sheep hook in paintings or photographs, the herder leaning on it as he gazes in the faraway distance at his sheep. Far more than a picturesque prop, the hook serves several practical purposes.

The herder can catch or stop a runaway sheep by placing the metal hook around a back leg. The hook can also be used to separate sheep as they bunch up in a branding or docking chute, or can simply serve as the herder's walking stick in his daily wanderings among the grazing sheep.

DOGS. Dogs have always played an important role in the sheep business of the West. The black and white Border collie is the dog most closely associated with sheepherding, although certainly other breeds have been used to work sheep. Widespread use of the Border collie as a herding dog in the United States began in the mid-1800s when Scotland exported these animals for use in such northern states as New York, Ohio, and Michigan. The western sheep dog descended from these early Scottish imports and participated in the west-to-east sheep drives of the late nineteenth century.

An innate herding instinct allows the Border collie to be easily trained by both humans and older working dogs among whom he is often raised. Whereas a cattle dog will nip at the heels of cows to keep them moving, this approach is too rough for sheep who must be worked in a slower, gentler fashion. According to Edward Wentworth, a "smart sheep dog was like a smart herder—he made the sheep think they were going the way they wished, rather than that they were being driven."[12] A herder

Herders used a sheep hook to gather wayward sheep by catching the animal's leg. (Photo courtesy Museum of the Rockies, Montana State University, John Haberstroh Collection)

A sheepherder with his dog. (Photo courtesy Museum of the Rockies, Montana State University)

needed only one dog but often kept two as companions for each other. The canine diet consisted of table scraps or perhaps some old mutton supplied specifically for dog food.

Trained to obey a series of hand commands and whistles, the dog had the job of keeping the sheep together and rounding up the inevitable stray at day's end when he helped bring the band to the nightly bedground. Although not officially a guard dog, the Border collie aided the herder in protecting the sheep from such predators as coyotes, wolves, mountain lions, and bears. It is not surprising that a strong bond often developed between the sheepherder and his dog due to the solo nature of the work. Worker by day, companion at night, the sheep dog exhibited a fierce loyalty to the sheep which were his charges and also to his master. Stories of sheep dogs that performed incredible acts of loyalty and bravery abound in western folklore, Shep being the most famous of all.

Shep won his place in history in 1936 at the Fort Benton, Montana depot where he watched the casket of his sheepherder master being loaded onto an eastbound train. Thereafter, Shep supposedly met every train that pulled into the station, ever alert for the return of his dead friend. Local railroad employees took pity on Shep, fed him and allowed him to take up permanent residency in a spot under the depot from which he could watch the trains.

In 1939 a Great Northern Railroad conductor, Ed Shields, wrote the story of Shep which appeared in newspapers throughout the world and earned the dog a feature in Ripley's "Believe It Or Not." Shep received letters from many people offering to adopt him, and "some fifty sheepherders...had asked to have Shep, often coming in person to see the dog and to make their request."[13]

Shep continued his vigil for five and a half years at the Fort Benton depot. In 1942 a train ran over the aging collie as he watched and waited at his familiar post. Shep received a Christian burial attended by hundreds. The railroad employees erected a concrete monument to Shep above his gravesite overlooking the depot. The monument still stands today as a tourist attraction in the small town of Fort Benton.

TIN DOG. The tin dog was a simple device made by the herder with a piece of wire and old tin cans. The wire was bent into a circular or oblong form and strung with perforated tin cans. Hung by the wagon door within easy reach, the herder could toss the tin dog in the direction of a nighttime predator. The rattling tin dog helped keep the coyotes at bay.

HORSES. Horses were essential to the job of sheep herding. Two horses could pull the sheepwagon at lower elevations, while it usually took four horses on the trail to the summer range. The herder also had a riding horse for daily use, especially in the summer range where the sheep wandered further than during winter.

One rancher recalled that some of the old-time herders dispensed with the saddle horses, believing that they were more trouble than they were worth. These herders always took a lunch with them and spent the day walking with the sheep as they grazed, returning to the wagon only in the evening. Those herders who did use horses usually returned to the wagon for lunch.[14]

This same rancher also used only geldings in his sheep ranching operation. A mare was a liability on open range where herds of wild horses also roamed. A wild stud could lure the mare away and it was very difficult to recover her. If the mare did return she might have a new colt in tow, an unwanted burden to the herder.[15]

CHAPTER EIGHT

The Basque

No book about sheepwagons or sheepherding in the United States would be complete without some mention of the prominent role played by the Basque. Although many ethnic groups participated in the western sheep business, the Basque are most closely associated with it, even by people who may know little else about its history.

A popular misconception is that all sheepherders hailed from the Basque country located in the Pyrenees Mountains between Spain and France. It is also falsely assumed that the Basque came to herding naturally, having practiced it in their native land. In fact, many "Bascos" came from towns or small cities and had never handled sheep before. Other herders had grown up on farms, but rarely did a family have more than a hundred sheep in their flock.[1] Even if a Basque had some previous experience with the animals, it was not under the transhumance system practiced in the western states that involved thousands of sheep, hundreds of trailing miles, predator control, and months of isolation.

Basque families and a herder visit on the summer range in the Big Horn Mountains. (Photo courtesy Johnson County Library, Buffalo, Wyoming)

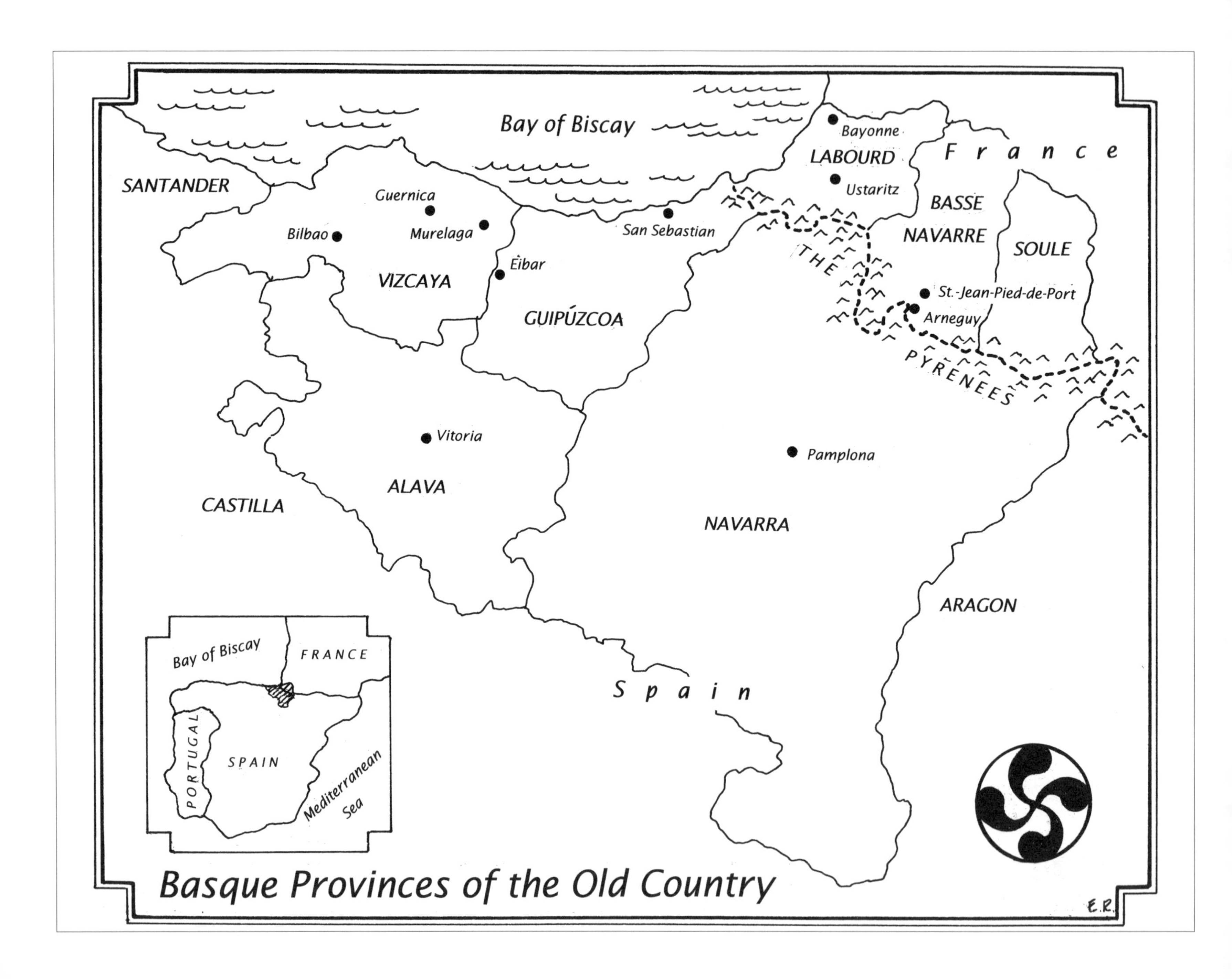

Basque Provinces of the Old Country

The Basque homeland is located in the Pyrenees Mountains on the Spanish/French border. (Illustration by Elizabeth Rosenberg)

The Basque left their homeland for the same reasons that millions of other ethnic groups did, to search for new economic opportunities and escape political unrest. The first large-scale Basque migration took place in the 1830s with Uruguay and Argentina as destinations, and it was in the treeless pampas that they first established themselves as sheepmen in a frontier situation. The California gold rush of the late 1840s and early 1850s lured some of these South American Basque north to the United States. Failing in their search for gold, the Basque immigrants often found employment in the sheep industry that was growing to meet demand for meat products in the gold camps.[2]

Direct migration to California from the seven Pyrenees provinces of the old country began in earnest during the 1880s as the Basque involvement in the sheep industry grew. Generally, the Basque did not intend to live permanently in the United States; rather, by hard work and frugal living, they hoped to eventually return home with enough money to get established there.

The typical Basque immigrant arrived in the American West with very little money and often found work as a sheepherder. Many herders took their pay in sheep, a system called "sheep on shares" in which a herder ran his sheep with those he tended and thereby built up a separate herd. By using this method, a herder could start his own outfit after four or five years. If he decided to stay in this country a while longer, he often recruited a family member or friend from the homeland to herd for him. This pattern of establishing a small flock, then bringing over a man from the old world to help, was to be repeated in all of the western states and is particularly associated with the Basque sheepmen.

According to Basque authority William A. Douglass, by the "1880s the Basques had established a reputation as the finest sheepmen in the American West, and by the first decade of the twentieth century a few Basques had become prominent sheep ranchers with impressive herds and private landholdings."[3] Basque herders became sought after by both Basque and non-Basque ranchers, and often replaced Mexican, Scottish, and Portuguese herders.

Douglass speculated as to why the Basque became the herders of choice. Perhaps a Basque herder had more incentive to take good care of a rancher's band of sheep if he was also running his own sheep with them. Because the herder hoped to

Six Esponda wagons leaving Esponda's Buffalo home headed to the summer range in the Big Horn Mountains, circa 1930s. John Esponda left the Basque homeland for California and eventually became a very successful sheep rancher in Buffalo, Wyoming. Esponda was knows as "King of the Basque" and brought many Basque men to Buffalo who began their careers in the sheep industry as herders. (Photo courtesy Johnson County Library, Buffalo, Wyoming)

have his own outfit some day, he was also more likely to learn all he could about a range sheep operation and be a more avid student of herding techniques.[4]

Douglass also dispelled the myths that the Basque took naturally to herding because they all had sheep in the homeland, and that they had a "unique psychological capacity for social isolation." Neither is true, Douglass claimed. "Rather, despite the romanticism surrounding them, they show a pronounced determination to undergo temporary physical and mental privation as an investment in a secure economic future."[5]

Many of these small-scale Basque sheep operations throughout the West became known as "itinerant bands" or "tramp herds," a derogatory term meaning the flockmaster owned no land of his own. Instead, he grazed his sheep only on the vast public lands. He and his sheep were constantly on the move, and often raised the ire of the so-called "legitimate" sheepmen by grazing on the same public ranges. A land-owning, tax-paying sheepman believed his right to the public grazing land took precedence over the nomad's. On the other hand, to the landless Basque sheep owner who might hope to eventually return to his homeland, owning land was a waste of money when the public land was theoretically free for anyone to use.

These attitudes caused conflict in the West, which was gradually resolved over a period of years as the federal government imposed stricter controls on grazing the public domain. By the mid-1930s, with the passage by Congress of the controversial Taylor Grazing Act, the tramp bands had become a thing of the past.

Not all Basque sheep operators were landless nomads, however. Some, like John Esponda of Buffalo, Wyoming, early on began to buy the land that became the foundation of vast sheep empires. Esponda arrived in the United States in the early 1900s. He joined his brother who had relocated to the less populated area of Wyoming after spending seventeen years as a herder and tramp sheepman on California's increasingly crowded public range.

After a few years as a herder, then foreman, for a large Scottish sheep outfit near Buffalo, Esponda formed partnerships with various men. Many were Basque immigrants from his native town of St. Etienne-de-Baigorry, whom he had brought to Wyoming to work in the growing sheep business. Esponda eventually went out on his own, all the while buying land, and by 1919 had become one of Wyoming's most prosperous sheep producers.[6]

Like many Basque men, Esponda visited the old country after becoming a success in America and there found a wife. The newlyweds returned to Wyoming and began their own family. Esponda became known locally as "King of the Basque" and his home became a center of Basque activities, especially on Sundays when his countrymen gathered to play *pelota*, a type of handball, in the outside court Esponda had built next to the barn.

Esponda is credited, both directly and indirectly, with the establishment of the Basque colony in Buffalo. His first partners became successful sheep ranchers and they in turn brought herders to Wyoming from the old country. Marriages among the families strengthened the ethnic identity. To Basque scholar William Douglass, the Basque typified the pattern of "chain migration," a process by which emigrants populate an area along kinship and old-world regional ties. It was a familiar pattern in most of the Basque towns throughout the West.[7]

Not surprisingly, the first generation of Basque men in America usually married Basque women. Like John Esponda, they might turn homeward for

Basque ranchers John Iberlin and his son Simon (foreground) trailing two sheepwagons to the summer range in the Big Horn Mountains near Buffalo, Wyoming, circa 1930s. (Photo courtesy Johnson County Library, Buffalo, Wyoming)

marriage partners, visiting the Basque country after five or so years in America. Some reunited with a wife or sweetheart who had waited for them, while others perhaps became acquainted with the sister or cousin of one of their compadres in America.

Many single Basque women also immigrated to the Basque enclaves of the western United States, usually following a brother or other relatives. They often found employment as domestics, cooks, seamstresses, or maids. Most quickly found a husband among the plentiful supply of single Basque herders. For many a Basque woman, a sheepwagon was the first home she shared with her new husband. The couple sometimes lived in the wagon for a year or two, eventually settling in town once children began to arrive.

In *The Basque Web*, a history of the Basque in Buffalo, Wyoming, author Dollie Iberlin described the early days of many of these unions and provided a rationale for the modest lifestyle still maintained by many successful Basque ranchers today. "The first Basque homes were usually sheepwagons...A barrel of water or a spring near the wagon serviced the honeymoon home. In the winter, sheep pelts, salted down and scraped, covered the floor and benches in wall-to-wall wool. Humble beginnings, but the Basque

Basque sheep rancher Gaston Irigary with his son Joe on the steps of a sheepwagon in the Big Horn Mountains, circa 1930s. The dishpan often hung on the bottom of the door when not in use. A box was often used as a step up to the wagon when it was parked. (Photo courtesy Johnson County Library, Buffalo, Wyoming)

ethnic background put great emphasis on physical endurance, and short-range comforts were not considered, only the long-range goal of acquiring land and sheep."[8]

Another institution that helped maintain the unique ethnic culture and identity was the Basque hotel, sometimes called a boarding house. These hotels were staples of any sizeable western sheep town with a Basque community. (Small towns like Buffalo, Wyoming had one Basque hotel; Boise, Idaho, a much larger city, had more than twenty.) Usually owned by a Basque couple, the husband often a former herder, the hotel provided the familiar tastes and sounds of the old country to thousands of newly-minted sheepherders. A Basque hotel in New York City might be the first stop in the United States for the Basque man heading west. Following a long train ride to his final destination, the future herder might stay at another Basque hotel until his relatives or friends came to fetch him.

The Basque hotel resembled what we think of today as a large boarding house. The owners and their children also lived there; the herders who spent many years at the hotel often became part of an extended family which included in-laws and friends

Basque ranchers, herders and children with sheepwagons in Buffalo, Wyoming, circa 1930s. No two wagons are exactly alike. (Photo courtesy Johnson County Library, Buffalo, Wyoming)

of the owners. Here one could hear and speak one's native language, eat the familiar Basque food served at every meal, play *mus*, the national card game, and catch up on news of friends and relatives in the Basque provinces from the newcomers. Letemendi's boarding house in Boise was full at Christmastime. "Boarders shared their beds with herders in from the hills for the holidays.... As many as fifty sometimes showed up for dinner. New Year's Eve always ended with snake dances from one boarding house to the next in the Basque enclave around Grove Street, and the parties and dances lasted all night."[9]

The owners of the Basque hotels operated as mediators between the old culture and the new, smoothing the way for the immigrant unfamiliar with American ways and language. They arranged for a herder's medical care and assisted with financial and legal affairs. Ranchers knew to inquire at the hotels when they needed to hire herders. Basque women often stayed in the hotels to await the birth of their children, with the female proprietor acting as midwife.[10]

One herder summed up what Letemendi's hotel in Boise, Idaho, meant to him. "Almost everything I became in America I owe to Letemendi's. I met my

Basque visiting on the trail, Big Horn Mountains, circa 1920s. More so than other ethnic group, the entire Basque family spent time on the summer range. (Photo courtesy Johnson County Library, Buffalo Wyoming)

future wife in their kitchen. I learned to speak English talking to Vicki Letemendi's children. I got my first job in America through the boarding house. They helped me send money to my mother. I don't think I could have made it here without them."[11]

In some cases, a Basque man might leave his wife and children behind in the homeland and was little more than a stranger to them when the family reunited. One Basque woman recalled that, as the first-born American child in the family following a separation of four years, she became the apple of her father's eye which caused resentment among the older siblings who hadn't known their father in their young childhood.[12]

The wives of the Basque ranchers generally took a more active role in the ranching business than did their contemporary, non-Basque counterparts. Certainly many ranch wives of all nationalities were involved in various cycles of the business, most remembering the nonstop cooking at shearing time. But a Basque woman usually spent all or part of her summer in the mountains, living in a sheepwagon along with her husband and children.

A number of non-Basque ranchers recall that their mothers did not spend time at the "sheep

Basque matriarch Catherine Marton spent every summer in the Big Horn Mountains. Marton was the wife of a prominent Basque rancher in Buffalo. Her wagon resembles the "Home on the Range" model and is on rubber-tired running gear, circa 1950. (Photo courtesy Johnson County Library, Buffalo, Wyoming)

Contemporary Basque ranchers John Camino and his son Peter John Camino. Peter John's sons are joining him as the fourth generation of Caminos in the sheep business in Buffalo, Wyoming. One of the Camino wagons is a rare gabled-roof model built in the 1950s. (Author's photo)

camps" during the summer months. The women stayed back in town at the family house, taking care of their daughters who weren't as directly involved in the sheep business as their brothers. Unlike cattle ranchers, most sheepmen and their families did not live at a "home" ranch; most of them lived in towns, perhaps for close contact with the local mercantiles from which they supplied their herders. The matriarch of the sheep ranching family might visit the mountain camp once or twice a season—or maybe not even that, for some ranchers believed the camps were no place for women.[13]

One can only speculate as to why the Basque women were more involved in the sheep business than their non-Basque counterparts. Perhaps it had something to do with the strong Basque matriarchal tradition. All Basque were outsiders in the United States and like any recent immigrant group, they tended to stick together and maintain their old-country traditions. Also, these versatile women, most recent immigrants to this country, often came from a background in which they worked a variety of jobs. Always known as hardworking and thrifty, Basque women in the sheep business helped increase their husband's profit margins.

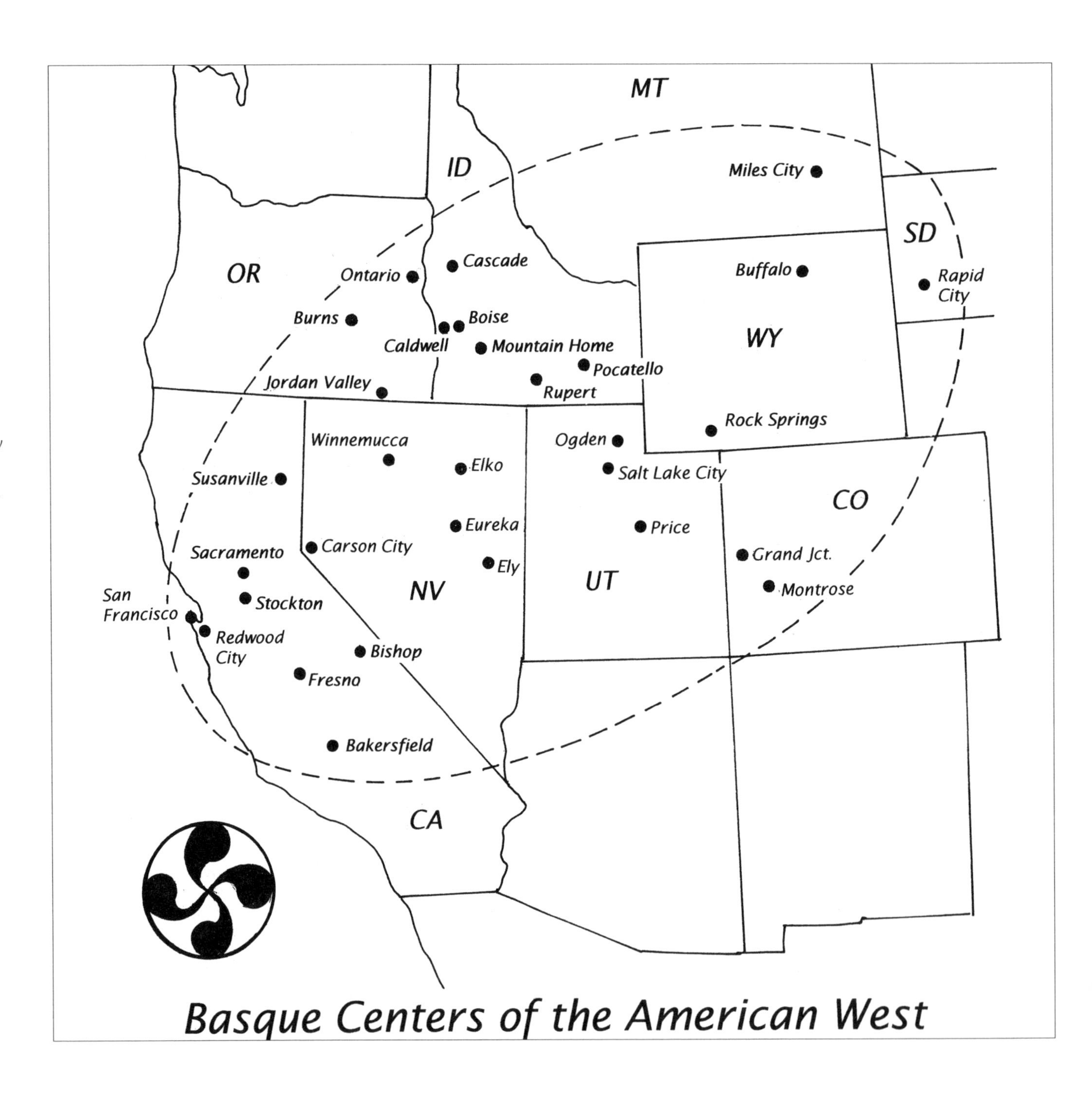

Map shows the contemporary Basque settlements in the American West. (Illustration by Elizabeth Rosenberg)

National Association of Basque Organizations sheepwagon parade, 1995 in Buffalo, Wyoming. More than thirty sheepwagons joined the parade. Some traditional wagons were horse-drawn. (Photos courtesy Richard Collier, Wyoming State Historic Preservation Office, Department of State Parks and Cultural Resources)

NABO sheepwagon parade, downtown Buffalo, Wyoming, 1995. Basque families from various western states rode in the parade. (Author's photo)

The long tradition of Basque herding is now a thing of the past. By mid-century, improved economic conditions in the native Basque country meant less incentive for men to seek work in the United States. As the American sheep industry waned and the economy in the western states diversified over time, the Basque found other employment in such industries as mining.

However, the Basque culture remains visibly alive today in the West. The National Association of Basque Organizations, known as NABO, holds a festival every summer, rotating the site among such Basque enclaves as Elko, Nevada; Boise, Idaho; and even small-town Buffalo, Wyoming with a population of less than four thousand. The three-day festival draws thousands of people and is highlighted by lively dance performances by the Basque clubs established throughout the West, ethnic food, and a mass conducted in the Basque language. It is a time for Basques to reunite, celebrate, and reconfirm their colorful heritage.

That Basque identify with the western sheep business and sheepwagons is clearly evident by these headstones in the Buffalo, Wyoming cemetery, 1996. (Author's photos)

Sheepwagons are an important element of that Basque heritage in America. The highlight of the 1995 NABO festival parade in Buffalo included thirty sheepwagons from such states as Wyoming, Montana, and Idaho. Freshly painted green wagons with new canvas or gleaming metal tops rolled down Main Street to the cheers of thousands of onlookers. The spectacle featured traditional horse-drawn wagons as well as those modified with a rubber-tired chassis. Proudly displayed on each wagon was the Basque name of the owner; family members, some dressed in colorful Basque garb, waved to the throng from the doors and windows of the wagons. The sheepwagon parade is still talked about today among those sheep-wagon aficionados lucky enough to have seen it.

In another eloquent tribute, the sheepwagon is featured on some Basque headstones. A number of these finely chiseled grave markers have a tableau with the same simple elements found in many sheepwagon paintings: a herder, sheep, a dog and a sheepwagon. These headstones testify to the strong identification of the Basque with the western sheep industry.

C H A P T E R N I N E

Women in Sheepwagons

Some enterprising women also lived in sheepwagons with their herder husbands and often a child or two. Their stories, which often included domestic details, add a fascinating dimension to the study of daily life in a sheepwagon.

Sedda Strickler Hemry

The Hemry family of Casper, Wyoming has contributed a remarkable visual legacy of sheepwagon life around 1900. Through photos he processed himself, Charles Hemry, an avid amateur photographer, documented the first year and a half of his married life, spent in a sheepwagon. He arrived in Natrona County, Wyoming in 1895 and had become a sheep rancher by 1901, when he returned to Ohio to marry schoolteacher Sedda Strickler, a childhood friend. Hemry brought Sedda to his sheepwagon home where they lived for over a year. Their first child, Wyoma, born in 1902, shared the sheepwagon with the young couple.[1]

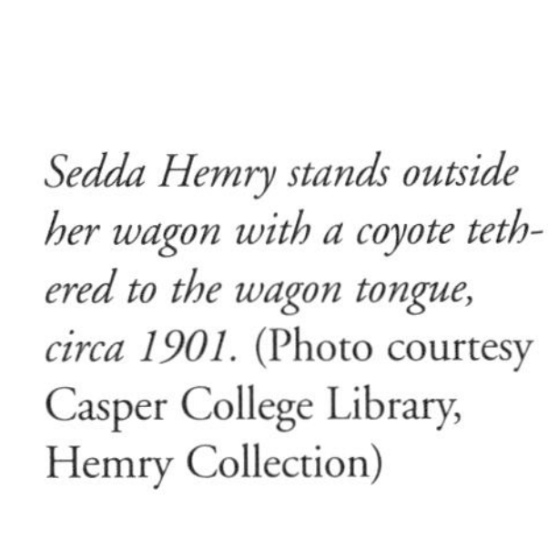

Sedda Hemry stands outside her wagon with a coyote tethered to the wagon tongue, circa 1901. (Photo courtesy Casper College Library, Hemry Collection)

Sedda's photographs reveal that she created a homey oasis in her sheepwagon on the lonely windswept

Above: Sedda and Charles Hemry's first child was born while they lived in a sheepwagon, circa 1902. Mrs. Hemry had a decorative runner on the steps leading into the wagon. (Photo courtesy Casper College Library, Hemry Collection)

Right: Sedda Strickler Hemry, 1901. She came to Wyoming from Ohio just days after she married sheep rancher Charles Hemry. (Photo courtesy Casper College Library, Hemry Collection)

range. Framed by a pair of long tie-back curtains, the bed is covered with a colorful quilt and satin pillows. Lace curtains hang at the window, a doily spread on the shelf above. Family photos adorn the patterned oilclothed walls and two shotguns rest in the gun rack. A runner covering the narrow floor space is visible in one photo.

Closeup shots of the tiny kitchen area also reflect Sedda's determination to make a comfortable, cozy home. China cups hang neatly from a bow; a coffee grinder is mounted above the flatware and food tins. A wedding gift of a silver creamer sits on the shelf above the stove.

Photos over a period of time reveal that just as one redecorates a house, Sedda changed the decor of the wagon—and perhaps even changed wagons since the back window appears different in some views. She rearranged photos and other sentimental pictures, hung a calendar, added more books and a baby's hairbrush. But the two guns remained, evidence of the sometimes hostile environment the young family shared with such predators as coyotes and mountain lions, and unwelcome strangers.

After more than a year, the growing Hemry household expanded into a one-room log cabin. The

Sedda's first home in Wyoming was a sheepwagon which she lived in for a year and a half. Sedda created a comfortable home for the couple inside the sheepwagon. A lace canopy separates the bed area from the rest of the wagon. A lace valance hangs above the three-paned window. This photo was taken in 1901. (Photo courtesy Casper College Library, Hemry Collection)

Baby Wyoma Hemry is posed in the door of the sheepwagon, circa 1902. Photo courtesy Casper College Library, Hemry Collection)

sheepwagon remained parked nearby, no doubt called into service as an extra room for friends and family visiting from "back East."

The Hemrys were not the only family living in a sheepwagon near the town of Wolton. Daughter Wyoma years later recalled "there were several frame homes, log houses, tents, and sheep wagons in this new town. About this time in Wolton, a mother and three young folks arrived: Mother, a 16 year-old daughter, a son about 13, and a girl about 11...They lived in a sheepwagon." [2]

Lucy and Nellie Morrison

Although books about the history of the western sheep business are few and far between, a noteworthy one, *Lady of a Legend*, was published in 1979. The authors, Wyoming natives Bob Edgar and Jack Turnell, tell the tale of Lucy Fellows Morrison Moore who became known as "the Sheep Queen of Wyoming."

Lucy Fellows was born in 1857 in a California gold camp. Her parents had left their native Vermont in 1852 and made an arduous wagon-train journey, which included three winter months camped near Casper, Wyoming, before reaching their destination, the gold fields of California. Barely eking out a living

Wyoma Hemry sleeps in the sheepwagon bed, 1902. (Photo courtesy Wyoming State Archives, Department of State Parks and Cultural Resources)

in the Golden State, by the mid-1860s the family relocated to Idaho Territory where gold had recently been discovered. It didn't take long for Lucy's father to give up the gold hunt and turn to farming at which he became fairly successful by supplying the Montana gold camps.

In 1873, sixteen-year-old Lucy married Luther Morrison, a family friend who had accompanied her parents on their westward trek in 1852. By 1880, Luther and Lucy were raising sheep on an increasingly crowded range in southern Idaho. Hearing stories of Wyoming Territory's Wind River Valley, a sparsely populated, semi-arid area of light snowfall and nearby mountains, Lucy and Luther, along with their three young children, left Idaho in 1882, trailing three thousand sheep eastward to their new home.

The Morrison family spent a frigid, snow-covered winter at Pacific Springs, a mining camp near the famed South Pass, and come spring their band was reduced to a mere two hundred sheep, half of them wethers (castrated males). After reaching the Wind River Valley, they lived in a tent for the next four years while they established a ranch and rebuilt their herd. Following the birth of their son in the tiny town of Lander in 1884, Lucy Morrison did not see another white woman for the next five years. Other than her family, the only people she had contact with during that period were trappers, Indians, an occasional cowboy, and soldiers from the fort near Lander.

Lucy's story, as recounted by her daughter-in-law Nellie in *Lady of a Legend,* is the stuff of pioneer legend. She helped her husband herd sheep, often with a baby on her hip. Until 1888 when the railroad reached Casper, the nearest wool shipping and supply point was 150 miles south, in Rawlins. Twice a year Luther made the three-week journey to town, leaving Lucy at home to take care of the sheep and the family, and live in fear of wild animals, raging blizzards, and Indians who wandered onto the ranch from the nearby Shoshone/Arapaho reservation.

Luther Morrison died in 1898, leaving the now large sheep ranch for Lucy and their only son, fourteen-year-old Lincoln, to operate. As more sheep and cattle moved into the Wind River Valley area, conflicts arose over an increasingly limited grazing range. In 1904, a lone gunman shot Lincoln as he stood at the door of a sheepwagon. Although her son recovered, Lucy remained bitter towards most cowmen for many years. The sheep/cattle wars continued to plague their operation until 1916 when Lucy hired two range detectives and the harassment stopped for good.

Well into her sixties, Lucy Morrison continued to work her sheep throughout the yearly cycle. She preferred life in a sheepwagon to the more civilized comforts of a house, even on her spread where the ranch hands and cook had permanent living quarters. She and her second husband also lived in the wagon. In later years she said, "I loved the nomadic life. I had looked forward to having a home, but one winter shut up in it left me with a deep distaste for what seems to me imprisonment behind four walls and the never-ending and boring demands of housekeeping."[3] Apparently, Lucy was no fan of cooking either; whether in a house or a sheepwagon, Lucy always left that chore to others and preferred a man cook.

As for Lucy's moniker "Sheep Queen of Wyoming," she explained, "It was not I who called myself the 'Sheep Queen.' It was first used sarcastically by men who envied me and wanted my range, even though I did not deliberately put any obstacles in their desire to acquire homes and herds...I did resist their efforts to push me off my own lands. In time newspapers began to me call me the 'Sheep Queen,' and people told me, out of respect for my hard-earned success."[4]

Lady of a Legend is actually the story of two women, Lucy and her daughter-in-law Nellie Morrison. Separated by a generation, their stories parallel the establishment and swift rise of the sheep industry in the large western states over a fifty-year period. Their tale also recounts the rapid industrial advances made in the West. For instance, they witnessed how quickly the railroad arrived and what an enormous change it made in ranching operations, greatly speeding access to the eastern markets on which the rancher was dependent. Likewise, the automobile began to transform all ranching operations almost as soon as it was introduced. As they did elsewhere, these advances sped up time and shortened distances on the range, perhaps most importantly for the women who now no longer had to endure their husband's semiannual three-week absences to purchase supplies.

Nellie weaves her own story into Lucy's account. By the time Nellie arrived in Wyoming in 1916, the Morrison sheep operation was prospering. Yet in some ways, after marrying Lincoln, this twenty-two-year-old city girl from Ohio faced the same travails

Lucy Morrison, the Sheep Queen of Wyoming, in later years. (Photo courtesy Casper College Library)

her intimidating mother-in-law had forty years earlier as a newcomer on the Wyoming range. Nellie's story speaks to the experience of the many women who found themselves, often through marriage, removed from the "civilized" environment of an eastern city or a small midwestern town to the often harsh and lonely landscapes of the large western states. Whether it was ranching sheep or cattle, the adjustments these women made to an utterly foreign way of life were remarkable.

Nellie's introduction to sheep ranching took place on the mountain summer range where Lucy and her second husband shared a sheepwagon. The young couple lived in a tent which Nellie described: "My own first home on the range was an eight by ten tent, furnished only with a bed and a big packing box for a dresser...the floor was virgin meadow grass...I consoled myself silently that housekeeping here would be simple."[5]

Imagine the young bride's disappointment when she realized that, just as her husband had warned her, there would be no use on the range for her steamer trunk and two suitcases filled with what Nellie considered "the absolute minimum of necessities, such as embroidered linens, even a flower vase and other such articles as useless in that tent as in an Indian teepee." Nellie was to quickly learn that "I too could pack a few essentials in a 'warbag'—a heavy seamless seed sack—and make out very well for a short time."[6]

That first fall, Nellie and Lincoln moved into a sheepwagon where they lived for the next year until Nellie had their first child. Many years later she recalled: "We cleaned and painted the inside of our sheepwagon. Years of housekeeping by sheepherders had left it dismal, dirty and scarred. I put up some bright hand-made calico curtains at the back and door windows. I bought a few china dishes to replace the chipped enamel ones. When I surveyed the whole new inside, I was glad my mother-in-law had already departed for California before she saw how I had 'gussied up' one of her sheepwagons."[7]

As many women can still recall, and most likely a number of men too, cooking in a sheepwagon took some getting used to. Nellie had vivid memories of her early culinary adventures on the range: "My knowledge of cooking was not extensive, but to get meals on that little wood stove was a real challenge to all the ingenuity I had, and almost defeated me. But I had adjusted to other primitive conditions recently, and decided I could to this. In time I came to love the cheer and comfort of that warm little stove when we came in from the bitter cold."[8]

Nellie's town-bred "ingenuity" included a practical though costly method of procuring that key staple of life on the range, bread:

> *After several failures on my part in trying to make edible yeast bread from the over-night sponge method, I gave up the effort during the extreme cold weather. My husband and the camptender teased me about the heavy gray coarse bread so much that I bought a bakery bread every time we went to town. Facetiously, they tried to convince me that my bread had much more 'body' to it than the light spongy bakery product. I can't imagine now why we had not bought bread before we did on our trips to town, except that at that time, it was an unthinkable, unheard-of extravagance to buy baker's bread for camps, or even ranches. If you ran out of bread, you made biscuits.*[9]

Nellie spent her first Wyoming Christmas in a sheepwagon which she adorned with "a few bright decorations for some juniper branches I had put up on the wagon bows." She cooked "a lamb roast for Christmas dinner, and with canned sweet potatoes, canned cranberry sauce, olives, fresh celery and canned plum pudding, we dined rather well for Christmas dinner in a sheepwagon far out on the storm-swept range."[10]

Like so many city folk, Nellie had sentimentally hoped for a white Christmas that first year, in spite of Lincoln's protests that "any stockman on the open range has no great longing for a white Christmas. We have learned that a blizzard around Christmas is apt to be followed by weeks of storms and below zero weather." Nellie got her wish and an adventure with it.

She later recalled, "My first Christmas day as the wife of a shepherd remains in my memory as the most vivid and exciting experience of those first months in Wyoming. In the gray dawn of that Christmas morning of 1916, a blast of wind and snow came roaring out of the north and shook the sheepwagon, startling me to full wakefulness. I stared out of the rear window over the bed, and it was already plastered with snow."[11]

The blizzard raged Christmas day and night and into the following morning, when Lincoln bundled up and left the sheepwagon in search of lost sheep.

Nellie, now alone for what would be two days and nights, recalled, "I just barely refrained from whining that never in all my life had I spent a single night alone, not even in the city, much less in such a great empty wilderness as this. But before I spoke, there came the fleeting thought that I was the girl who for years had wanted adventure in the highly advertised great open spaces. Well, this was it, and here I was in the midst of it!"[12]

Nellie emerged from her baptism by snow with a new-found pride in her endurance and an appreciation that "this often wild and savage land still demanded the same sort of courage, strength and endurance that it had of the early pioneers."[13]

Adelaide Hook Gilmore

Fortunately for us, a number of women kept diaries of life on the early sheep range. In her memoirs, Adelaide Gilmore drew on a detailed diary she wrote during her first eight years of marriage to a Wyoming sheepman.[14]

During their two-year engagement, Adelaide's fiancé, Frank Gilmore, made a six-month journey trailing his first flock of sheep from Idaho to Wyoming, where he established himself in the sheep business near Cody. Before departing to California for his wedding to Adelaide in 1897, Frank had ordered a "sheep camp wagon" as a more comfortable home for his new bride than the tent that had served him the past two years.

Adelaide described this sheepwagon, built in Billings, Montana:

> *The camp wagon was built on a regular Studebaker farm wagon body. Side boards had been built out along the sides—with a storage box in the center of each side, and down on the outside, opening with a lid inside. The back end had a platform built out for a bed and underneath, in the wagon bed, a storage place. Over this "living part" of the wagon was a curved heavy canvas cover. In the center of the front end was a post, from the bed of the wagon to the top of the canvas cover. On the right side, looking out, the canvas was left loosed to be tied and untied at the post and that was our door. At the left was the stove and back of the stove, on the side, as a built in cupboard. Not many "comforts of home" but we managed to live and enjoy a primitive way of life.*

Adelaide Hook Gilmore in front of a sheepwagon near Cody, Wyoming, circa 1899. (Photo courtesy Park County Historical Society Archives)

She also observes that their wagon was "one of the first brought into that part of the Big Horn Basin."[15]

From Adelaide's account, it's clear that the familiar Dutch door, a key characteristic of the sheepwagon as we know it, was not a standard item in 1897. However, by 1903, the Studebaker Wagon Company featured the double door in their catalog. Throughout her narrative, Adelaide referred to their "sheep camp wagon" and "camp wagon," adding to the list of names by which the wagon was colloquially known. Adelaide also remembered seeing that first cousin of the sheepwagon, the cooster wagon.

In the isolated and often hostile environment of the sheep range, Adelaide also felt the loneliness so many pioneer women experienced due to the lack of female companionship. She recalled, "It was weeks before I saw a woman again." Even by the standards of other frontier women, life in a sheepwagon was viewed as a real hardship. Adelaide recalled, "The few women I would meet that winter always asked me how I could stand it—living in a sheep wagon. I guess they could see that it was not hurting me."[16]

Adelaide's diary tells of the practical realities of living in a wagon, dealing with weather and food. She took note of the "intense cold...I had to learn to

Louise Turk and her step-daughter Vivian Turk in the Big Horn Mountains summer range, 1941. This was Louise's first summer in a sheepwagon. Louise has written a book about her life titled Sheep! Autobiography of Louise Turk, Woman Sheepherder *published in 2001.* (Photo courtesy Louise Turk collection)

be careful not to get my face, ears, hands and feet frozen when outdoors. To learn to sleep with our arms and heads under the covers at night when the fire was out in the stove, was very hard for me to get accustomed to. And to get used to having the water in your eyes and nose freeze and not be shocked when the men came in with icicles hanging from their noses and mustaches." A thermometer hung outside the wagon and Adelaide noted a record-breaking temperature of 54 degrees below zero during a nighttime blizzard in 1899.[17]

The Gilmore diet was exceptional compared to the standard sheepherder fare:

> *Most of our food was canned except our meat, bread and fruit. Our meat was mostly antelope, which was plentiful at that time, with once in a while a mutton. Our fruit was dried apples, pears, peaches, apricots and prunes which were washed, soaked and stewed. And our bread was always the "sour dough." As soon as I started living in the wagon I had to learn how to make sour dough bread and, believe me, it was good*

bread—the bread of the camps everywhere. We made our "starter" which had to get sour and "alive." We cared for this like a baby and it went on indefinitely. When making our bread we always left a little of the starter for the next batch of bread.[18]

The dried fruit Adelaide spoke of was probably a luxury not afforded most sheepherders. She tells of her father-in-law, at home in California's fertile Central Valley, sending sacks of dried fruit and almonds and peanuts every fall. "I always had a pan of roasted peanuts for our evening 'nibblings.'" The Gilmores also supplemented their diet with butter, eggs, and chicken from ranchers they occasionally visited. "We would dress the chickens, hang them up over night, and they would be frozen by morning, and they would keep in that condition until we thawed them out to cook, as we wanted them. We didn't need any 'deep freezer' or refrigeration."[19]

Adelaide's memoirs are so chock-full of insights into life on the sheep range a hundred years ago, one can only wish she had published the entire diary she kept while living in the sheepwagon. In 1905, after eight years in Wyoming, the Gilmore family returned to California where they purchased Frank's father's ranch and began a new chapter in their lives.

Louise Turk

Born in 1921, Louise Brown Turk grew up near Sussex, Wyoming during the Depression of the 1930s. Louise remembered playing in a sheepwagon as a child and wanting to live in one. Her wish came true when she married Brookie Turk in 1940. Brookie worked for a nearby ranch as a camptender and sheepherder, and he and Louise spent many years together living in a sheepwagon.[20]

During World War II, the Turks abandoned their sheepherding life and followed Louise's family to Seattle, where Brookie worked in the defense industry making "big money" compared to his ranch wages of twenty-five dollars a month in 1940 Wyoming. After living with the Brown parents for a while, the Turks decided to build "a small shack" of their own. Louise desperately missed Wyoming and her life in a sheepwagon. She recalled their attempt to recreate at least part of that lifestyle in dreary, rain-soaked Washington state:

Someone nearby had torn down a building, so we bought enough scrap material to build

Louise and Brookie's son Pete takes a bath on the summer range in the Big Horn Mountains, circa 1945.
(Photo courtesy Louse Turk collection)

our "house"...When we finished building, we had something that was a cross between a sheepwagon and a present-day camp trailer without wheels. One end had a built-in bed with a table that pulled out from under it, just like a sheepwagon...A big bin with a hinged lid served as seating on one side of the table, and Brookie built a bench for the other side. At the other end of the house were bunk beds for the kids, with closets at each end and drawers under the bottom bunk where they could keep their personal things...We cooked and heated with a sheepwagon stove we'd had Harvey ship from Wyoming...We didn't have much room but I was happier than I'd been anywhere else we'd lived, probably because it seemed more like my wagon home in Wyoming." [21]

As ranch workers joined the armed forces or fled to the big cities for high-paying defense industry jobs during World War II, many ranchers were left without experienced help back home. Top hands like Brookie Turk, once they became employed in the war industries, were not allowed to quit or they would be immediately drafted into the army. This

Brookie, Louise, and Vivian Turk at the summer sheep camp in the Big Horn Mountains, circa 1941. (Photo courtesy Louise Turk collection)

Pete and Louise Turk outside their sheepwagon, 1945. (Photo courtesy Louise Turk collection)

ban was lifted mid-war due in part to the agricultural labor crisis, and Brookie and Louise happily returned to Wyoming in August 1944.

The Turks spent the following winter, with their new baby, in a sheepwagon parked near the Meike ranch in Sussex. Each summer they trailed a band of sheep to the Big Horn Mountains where a sheepwagon housed them and their three young children. For his family's comfort, Brookie built a series of outhouses on the various camp grounds of the summer range; these still stand today. Louise and the children, and many of their friends and relatives, have fond memories of those long summer days in the mountains beside a sheepwagon. Louise never liked returning to their house for the winter where, she said, "I hated housework!" [22]

Louise not only helped herd sheep but was the main cook for what could be quite a crowd at shearing time. Louise recalled a time on the trail to the summer range when she cooked three squares a day for some twenty-three men on the little sheepwagon stove. Rising at four A.M., she prepared eggs, bacon, potatoes, bread, and coffee for breakfast; a large dinner at noon consisting of roast meat or stew, potatoes,

Louise Turk, 1995 (Author's photo)

perhaps some green vegetable like peas, more bread, pies, and coffee again; and an evening supper, usually sandwiches made from the leftover meat, coffee, and perhaps pie or pudding for dessert. The men ate in shifts which meant when she was not cooking or serving, she was busy washing dishes and planning the next meal. She worked longer hours than the men, finishing at ten P.M. after breakfast preparations were made. Louise, who didn't know how to boil water when she married Brookie Turk, quickly learned to feed a small army.[23]

Another woman recalled a major culinary feat when she cooked for a large crew with the help of a woman friend. They joined two sheepwagons together by placing a board horizontally from door to door that allowed them to use both stoves in close proximity. Women and men who lived in sheepwagons learned to be inventive.[24]

CHAPTER TEN

Sheepherder Superstitions, Customs, and Pastimes

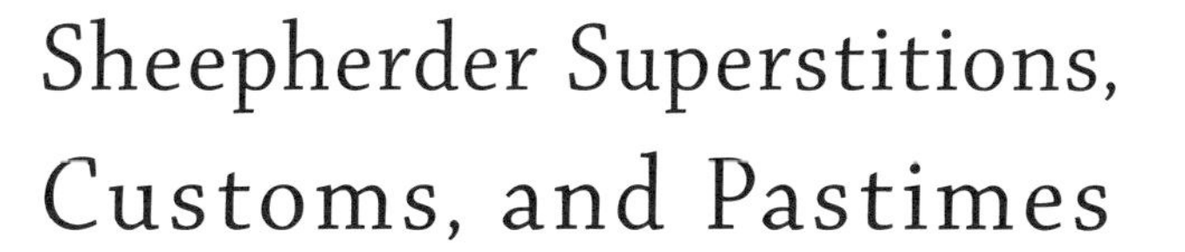

A NUMBER OF RANCHERS remember superstitions that developed among the early herders. Some herders refused to stay in the sheepwagon if the door faced any direction other than southeast. If the wagon was parked with the door at the southeast, then one always slept with one's head to the northeast. If the feet faced north, rather than the head, it was a sign of death and burial.

Practical factors also dictated the placement of the wagon. The door usually faced southeast because the prevailing winds blew from the northwest, and the morning sun streaming through the door's window would help to heat the wagon. Meat was often stored on the north side of the wagon where it stayed cool. Sourdough starter placed near the stove remained warm. Bacon grease rubbed on the end of the wagon tongue protected the wagon from violent lightning storms on the summer range. As the herder set the wagon in its resting spot at camp, he might place a cup of water on the table to ensure the wagon was properly level on the ground.[1]

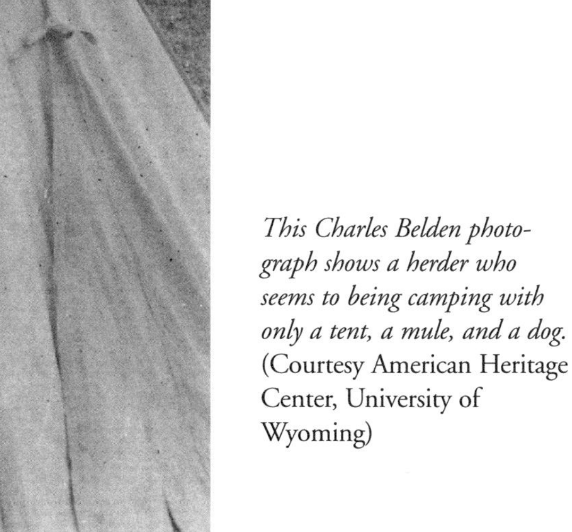

This Charles Belden photograph shows a herder who seems to being camping with only a tent, a mule, and a dog. (Courtesy American Heritage Center, University of Wyoming)

Sheepherder carvings on aspen tree near the Colorado/Wyoming border. The carving in the left photo marked the sheep trail, while the carving in the right photo reflects a lone herder's fantasy. (Photos courtesy Richard Collier, Wyoming State Historic Preservation Office, Department of State Parks and Cultural Resources)

Besides reading or perhaps making music with some type of instrument, herders amused themselves in a variety of ways during the slower summer months when being outside was pleasurable. Hunting for arrowheads occupied many hours, and the cache gathered on the summer range could be sold in bars during the herder's annual vacation toot in town.

Other produced drawings on tree trunks, often referred to as "aspen carvings." Many of these images can still be seen today in the Medicine Bow Mountains along the Wyoming-Colorado border.

Herders also spent time building what are referred to as "stone johnnies" or sheepherder monuments. These piles of stone laid one on top of another, with no mortar in between, might be built as high as twelve feet. The stone monuments are found all over the former sheep range of the western states.

Often visible from miles away, they may have served purposes other than merely killing time. Simply functioning as markers on the open range could be useful. One herder credited the stone piles with saving his life, acting as guideposts to his camp during a winter storm. Names were given to some of the monuments: "Dinner Table Monument," a lunch spot for the herder; "La Mesa Monument" where a flat, table-like stone topped a rock pile. On a sage-covered range, barren of any trees, the monuments also served as sentinels for marking traditional camp sites and grazing areas for the herder, places he might return to year after year.

The more distant origins of the stone monuments are a source of mystery to some. An old Wyoming herder, writing in the 1930s, recalled that as a boy herding his father's sheep, he and others had built similar monuments in their native South America. He also saw manmade stone cairns when herding in Mexico, and speculated that the practice traveled to the western United States with the herders from "old Mexico." Always curious about the custom of building the stone piles in sheep grazing areas, he was told by older herders, "There have been herders' monuments in this country as long as I can remember." [2]

Some sheepherders composed poetry out on the range, although very little of it was apparently published or recorded. A 1947 magazine article quoted verse from an unknown herder:

I've heard men long for a palace;
I want no such abode,
For wealth is a source of trouble and
a jeweled crown is a load.
I'll take my home in the open
with a mixture of sun and rain,
So give me my old sheep wagon
on the boundless Wyoming plain.[3]

A sheepherder monument in northwestern South Dakota, also called a "stone johnnie." Herders built these monuments as markers on the landscape and to pass the time. (Author's photo)

The poem that follows appeared in *The Wyoming Stockman-Farmer* in August 1920:[4]

A Sheepherder's Monument

Silent pile upon the hilltop,
With the vastness all around;
Landmark of the lonesome shepherd,
Set 'mid silence most profound.
Guide for men and flock returning
O'er the hills near daylight's end.
Man who has no other helper
Than his faithful canine friend.
Monument of patient labor,
Builded slowly stone on stone;
Built to while away the hours,
Hours he spends there all alone.
Who can tell what thoughts come o'er him,
As he worked there all alone!
Some past tie he may have severed,
With the placing of each stone.
Each stone may be a hope departed,
Or plan, that now lies buried deep,
While memories crowd in upon him,
In close array like flocks of sheep.
This pile of rocks may not be classic,
It may not be a work of art;
But it may be the sign of longing
In some poor shepherd's aching heart.

— T. S. PARSONS

Sheepherder postcard, circa 1917. This wagon is not the traditional form of a sheep-wagon; rather it resembles a tarpaper shack on wheels. (Author's collection)

C H A P T E R E L E V E N

Sheepwagons in War

IF PEOPLE KNOW ONLY one thing about the western sheep business, chances are it concerns the sheep/ cattle wars of the late nineteenth and early twentieth centuries. A fascinating and violent chapter in the story of western settlement, these range conflicts are still discussed only in hushed terms in those communities where the last incident occurred in the 1920s. Families of the antagonists remain in the area and the painful history, if transmitted at all, is told orally, not written down for public view.

The conflict was basically a battle over the free grass available on the public domain. Large-scale cattle herds generally preceded sheep by about ten years in the western states. Ranchers with thousands of cows dominated the range and turned their herds out to graze over the hundreds of square miles of public land for months at a time, sorting their brands out from others at roundup time. Conflict was minimal among cattle people, and sheep were tolerated as long as enough open range was available for all to operate as they chose. But once the range became

Sometimes it seemed to cattlemen that sheep would fill the range. (Photo courtesy Museum of the Rockies, Montana State University)

crowded and overgrazed, tensions arose between the two groups, with the cattlemen believing in the first-come, first-served principle, and disregarding the fact that the land was public.

The first round of confrontations occurred in the 1870s and 1880s in the southwestern territories of Texas, Arizona, and New Mexico which were settled before the northern Rocky Mountain states. From the 1890s until 1921, the conflict raged in the states of Colorado, Wyoming, Montana, and Oregon. All told, during the fifty-year war period, 128 violent incidents occurred; over 50,000 sheep were killed; 28 sheepmen and 16 cattlemen died.[1]

The epithets used against herders were first heard about sheep: they were dirty, smelly, stupid, timid, and naturally inferior to cattle (or cowboys). One of the cattlemen's most effective pieces of propaganda, still believed by some today, was that sheep ruined an area for cattle due to their grazing habits and the offensive scent they left on the range.

It is true that if overcrowded or held too long in one area, sheep crop the crowns of the grass and leave behind a pasture of little more than mud or dust. Sheep also secrete a sticky, odorous substance from a gland between their toes which cattlemen believed repulsed cattle from grazing the same area. Although untrue, this myth died hard, and not until cattle ranchers began to raise sheep along with cows did it begin to fade.

In fact, the two species worked well together in the large open spaces. According to David Dickie, a large sheep operator in western Wyoming from 1890 until 1935, "when his cattle grazed the valleys and lower slopes, and his sheep ranged the timbered areas and the hill or mountain tops, approximately a 20 per cent greater return was realized than when either species ran alone."[2]

The word "deadline" took on new meaning during this time. As the range conflict escalated, deadlines became an important weapon in the cattlemen's arsenal. A line, arbitrarily selected by the local cattle interests, would be drawn in the dirt marking the boundary between sheep and cattle grazing. The crossing of that line by sheepmen, whether intentional or not, signaled defiance to the cattlemen that justified a raid on the errant sheep camp.

A raid might involve poisoning the ground where sheep grazed with saltpeter (deadly to sheep but not cattle); lacing the feed with strychnine; shooting, clubbing, or dynamiting flocks; and "rim-rocking"

This raid on a sheep camp took place on April 24, 1904, twenty-four miles south of Laramie, Wyoming. Sixteen masked men set fire to two sheepwagons and clubbed three hundred sheep to death. Three herders were tied to the rail fence and warned not to cause any trouble or they would be killed. Losses totaled two thousand dollars; although four men were arrested, they were found not guilty.

"It was extremely difficult to prove the guilt of assailants in a sheep camp raid, when a skillful lawyer scrambled details from witnesses who were attacked in the dark by men using masks to cover their identity."

— Edward Wentworth, America's Sheep Trails. (Photo courtesy American Heritage Center, University of Wyoming)

A general view of the sheep camp after the Stevens-Maxwell raid. Men pull wool from dead sheep in the foreground, May 1, 1904. (Photo courtesy American Heritage Center, University of Wyoming)

—driving sheep over a cliff. One of the cruelest tactics was setting fire to the hapless sheep, and occasionally to the sheep dogs. Parts of the wagon, like the wheel spokes, might be used as clubs to beat the sheep and herder.

Other times, the sheepwagon, along with the supply wagon, was set afire with the herder tied up inside. More often, the herder was blindfolded, tossed out of the wagon and threatened by the masked intruders, but spared his life. In at least one instance, the sheep dogs were tied to the wheels of a sheepwagon that was then set on fire. Other times, the raiders shot the herder's horses.[3] In a particularly harrowing incident near the Wyoming–Colorado border, the tongue of the sheepwagon was propped up vertically with a noose hanging from its end. The herder was left hanging to die unless he agreed to "getting his sheep clear out of the country." The cattlemen were kind enough to lower the herder as he neared death and offer him the opportunity to save his life, which he took.[4]

Many states saw infamous raids that by now are more than just history. They have become folklore. Arizona's Pleasant Valley War of 1887 claimed nineteen lives of both sheepmen and cattlemen. In Idaho, the 1895-96 range conflict in southern Cassia County and across the border into Nevada involved a classic western-type villain named Diamondfield Jack, a sheep-hating zealot employed by cattlemen to protect a deadline. Colorado had ongoing conflicts, first in the south along the New Mexico border and later with the Mormon flockmasters on the state's northwest Utah border. One of the last incidents in the West's fifty-year range war took place in Routt County, Colorado in 1920.[5]

But it is the state of Wyoming that is perhaps most famous for violent sheep/cattle conflicts, which totaled thirty-five, more than in any other state. Notorious range rider Tom Horn, hired by such powerful cattlemen as the Wyoming Stock Growers Association and Swan Land and Cattle Company to patrol rustling, also had a role in the sheep wars. Convicted of the murder of thirteen-year-old Willie Nickell, whose father had peppered the formerly all-cattle range with sheep, Horn was hung in Cheyenne in 1903.

Also well-known is the Ten Sleep Raid (sometimes called the Spring Creek Raid) of 1909 in Wyoming's central Big Horn Basin where two sheepmen and one herder were murdered. The masked marauders shot a French herder named Jules "Joe" Lazier and sheep rancher Joe Emge, before the two men had a chance to leave their wagon. This sheepwagon, along with another in the camp, and a supply wagon were set on fire. The next day, local law enforcement officials, accompanied by the county attorney and coroner, found the grotesquely charred bodies of the two men, and the corpse of a third man, Joe Allemand, who had been shot outside the wagon. Around them lay the two smoldering sheepwagons and dead sheep and dogs.

Wyoming newspapers followed every word of the sensational trial that followed in the town of Basin. The Wyoming Stock Growers Association reportedly raised $200,000 for the legal defense of the seven men arrested for the vicious raid, five of whom served time in the state penitentiary at Rawlins. A court case ensued between the governments of France and the United States over the death of French herder Joe Lazier. His family in France reportedly received $25,000 as settlement for his cruel death in a sheepwagon on a desolate stretch of Wyoming landscape.[6]

Although violence flared between the warring factions until the early 1920s, the West had become a different place by then; the vigilante era of "taking justice into your own hands" had passed. Growing public antipathy against the range skirmishes, along with the establishment of the Forest Service in 1905, presaged the end of the fifty-year conflict. The federal government gradually took control of the public lands from the ranchers' hands by issuing grazing permits that limited the number of sheep and cattle allowed to feed on the formerly open range. The 1934 Taylor Grazing Act protected the land from overgrazing and also eliminated those people known as tramp sheepmen, men who owned no land of their own and only grazed on the public land, a long-time annoyance to both cattle and sheep interests.[7]

Cattleman-Sheepmen War Statistics

	Violent Incidents	Sheep Killed	Sheepmen Killed	Cattlemen Killed
Wyoming	35	16,305	10	1
Texas	29	3,215	4	2
Arizona	23	5,420	11	13
Colorado	15	13,750		
Oregon	10	10,064		
Idaho	5		2	
Washington	5	300		
Montana	3	3,500		
New Mexico	3	700	1	
TOTAL	**128**	**53,254**	**28**	**16**

Source: Data from Bill O'Neal, *Cattlemen vs Sheepherders: Five Decades of Violence In the West, 1880-1920*, (Austin: Eakin Press, 1989), p. 16

CHAPTER TWELVE

Modern-Day Sheepherders

SUCH FACTORS AS THE widespread use of synthetic fabrics, central heating, decreased lamb consumption, competition from New Zealand and Australian sheep products, and the controversy over predator control have contributed to the ongoing demise of the United States sheep business in the last quarter of the 1900s. The sharp decline in the sheep industry has meant a subsequent drop in the number of herders and sheepwagons. However, the first signs of change in the western sheep business began much earlier, following World War II.

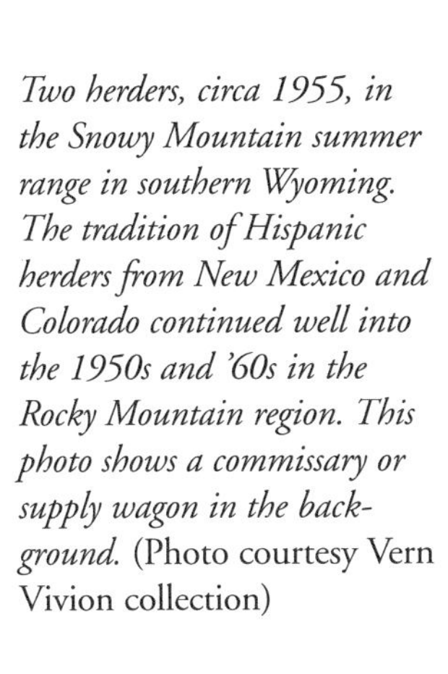

Two herders, circa 1955, in the Snowy Mountain summer range in southern Wyoming. The tradition of Hispanic herders from New Mexico and Colorado continued well into the 1950s and '60s in the Rocky Mountain region. This photo shows a commissary or supply wagon in the background. (Photo courtesy Vern Vivion collection)

Thousands of experienced agricultural workers never returned to their ranch or farm jobs after joining the military or finding high-paying employment in such defense industries as shipyards and airplane factories. Older ranchers date their continual lack of good help to the postwar years. As all wars do, World War II, and the prosperous decade of the 1950s that followed, created tremendous social shifts. A variety of job opportunities became available, and sheepherding, with its solitude and relatively low pay, held

Louise and Brookie Turk in the Big Horn Mountains of Wyoming. Louise became a sheepherder when she married herder Brookie Turk in 1941. Although rare, women did become sheepherders. After Brookie's death, Louise continued to herd sheep alone until 1997. (Louise Turk collection)

less appeal to the returning G.I. and the temporary war worker, American-born and immigrant alike.

Wartime innovations also changed ranch life. Surplus army jeeps, those four-wheel-drive vehicles first developed for the military, began to appear on western ranches, reducing a two-day horse trek to the mountains to a matter of hours. Ranchers began to fence large tracts of their deeded land to contain their sheep, and were able to check on them daily with the use of the jeep. In some cases, this gradually led to the elimination of herders and sheepwagons.[1]

The introduction of the four-wheel-drive vehicle also changed the sheepwagon itself. The pickup truck became the vehicle of choice to pull the wagon between the summer and winter grazing areas, reducing the need for horses. A rubber-tired vehicle towing a wooden-wheeled wagon proved inefficient because any speed over ten miles an hour caused damage to the running gear by overheating the iron wheel rims. Sheepwagon builders and ranchers solved this problem by converting their wagons to rubber-tired chassis. This evolution took place primarily during the 1950s and early 1960s.

The relatively few herders on the western range today still live in sheepwagons; virtually all of the

Peruvian herder in southwestern Wyoming, 1994. Everything has its place in a sheepwagon. The broom often hangs in this exterior position. (Author's photo)

A large number of modern herders come from South America. The wagon in this 1994 photo taken in Rock Springs, Wyoming has a small table attached to the bottom half of the door. An old tradition continues with the rubber skirting around the bottom of the wagon which provides additional insulation and protection for the dogs. (Author's Photo)

Charles Hemry, Wyoming winter range, circa 1901. Note the canvas skirting around the running gear. Compare this with the modern rubber skirting that serves the same function in the photo to the left. (Photo courtesy Casper College Library, Hemry Collection)

wagons, no matter how old or new, have rubber tires. The longtime tradition of the foreign-born herder continues but the countries of origin have changed. Since the 1970s, Mexico, Peru, and Chile supply most of the herders in the western United States. Recruiting is no longer done directly by the rancher but instead is primarily the job of three organizations in the West: the Western Range Association in California; Wyoming's Mountain Plains Agricultural Services; and the Wasatch Agency of Salt Lake City.

Ranchers pay membership dues to these organizations which in return supply the herders, often on three-year contracts. Perhaps their most valuable service is complying with the increasingly bureaucratic processes of the United States Department of Labor, and with the Immigration Reform and Control Act of 1986 that regulates foreign workers. In addition to handling most of the governmental red tape, these companies screen all herder applicants, arrange for their transportation to the United States, communicate in Spanish, and update members on changing regulations.

The herder's monthly salary is based upon a rate set by the Department of Labor of each state. In Wyoming, a herder in the year 2000 made about

A father-and-son herding team from Mexico stand at their sheepwagon near Opal, Wyoming in 1997. The rancher added a modern solar panel and rubber skirting to the wagon. (Author's photo)

Chilean herder Evangelisto Gonzales, shown in this 1996 photo, came to Wyoming's Warren Livestock Company on a three-year contract which is typical of the herding business today. Most of these men from Central and South America do not speak English. Many employment requirements and benefits have changed, but housing in a sheepwagon remains the same. (Photo courtesy Richard Collier, Wyoming State Historic Preservation Office, Department of State Parks and Cultural Resources)

$650 a month, while in California and Oregon the wages were higher. The herder is eligible for workers compensation but must pay for his health insurance and the federal income tax on his wages. A two-week vacation is provided but the herder can choose to take wages instead. The agencies pay up front for the herder's transportation to the United States, but it is up to the rancher to withhold the travel cost from the herder's wages and pay it back to the agency.

In traditional fashion, the herder still receives room and board as part of his wages, most often in a sheepwagon resupplied every week or so, as in days of yore, by a camptender or the rancher. Federal regulations also govern today's sheepwagons, which can be inspected at any time. Many ranchers have modified their old sheepwagons to conform to current OSHA regulations that specify the rear window opening must measure at least two feet by two feet to provide a second egress in case of fire or other emergencies.[2]

Another tradition continues today in that foreign-born herders often recruit male relatives or friends from the home country to join them in the western United States. This word-of-mouth method works to the advantage of both the rancher and the agency as the potential herder can learn the job, in his native

A contemporary Wyoming postcard shows two herders and a sheepwagon. (Photo by Beryl M. Williams. Reprinted by permission of Wyoming Book & Card, Mills, Wyoming)

tongue, from an experienced countryman. Such a relationship also helps to mitigate the loneliness and isolation the foreign-born herder experiences in a new country run by English-speaking strangers.

The old ways merge with the modern. A man accepts the job of sheepherding as a means to an end, as an avenue to a better life for his family. But now he often communicates with them not by letter but by tape-recording his voice as he tells of his new life in a strange country. The sheepwagon still provides a home for the herder, although solar panels might help to heat it and a cell phone may provide more immediate contact with his employer and the outside world. In spite of these conveniences, his traditional solitary lifestyle and his close bond with the elemental forces of nature and animals continues.

This rubber-tired sheepwagon, being pulled down a highway into the setting sun, may be a metaphor for the future of the sheepwagon. (Author's photo)

PART THREE

THE LEGACY

Inventions[1]

We sheepherders tried to modernize
And stay current with OSHA rules,
Cause riding wagons isn't safe
Behind a pair of mules.

So I removed the wagon tongue
And mounted it on the rear,
It's easier for the mules to push
And easier for me to steer.

The sheepwagon is very small
So tried some exploratory
And constructed a tandem cart
And a convenient second story.

I invented a special tub
To bathe in a lake or stream,
It has no visible bottom
Which makes it easy to clean.

The herders' catching cane is obsolete,
We simplified the catching trip
For distance is no problem
With a 20 foot velcro strip.

— Obediah Brown

CHAPTER THIRTEEN

Other Uses of Sheepwagons

THE VERSATILE, UTILITARIAN sheepwagon provided temporary shelter for a variety of uses as new towns sprang up almost overnight near the railroad and the burgeoning oil fields. For two years, a sheepwagon served as the schoolhouse in the oil boomtown of Grass Creek, Wyoming beginning in 1915. The younger pupils attended classes in the morning, and the older ones in the afternoon.[2]

Sheepwagons also housed a number of schoolteachers. Jessie A. Bryant recalled that as a baby she lived in a sheepwagon with her mother, who taught school near Saratoga, Wyoming.[3] Zella Pelloux Slayton of Buffalo, Wyoming found employment as a schoolteacher in rural Johnson County in 1940. Shortly after she began her new job, the teacherage burned to the ground. Zella purchased a sheepwagon for one dollar from her father, sheep rancher Martin Pelloux, and she lived in it for the next two and a half years while she taught. Similarly, Pauline Peyton agreed to become the teacher at the rural North Point School in Converse County, Wyoming "if she had a sheepwagon." A local

Zella Pelloux Slayton lived in this sheepwagon in Johnson County, Wyoming for two years when she taught at a one-room schoolhouse, circa 1940. Zella's father, a sheep rancher, sold her the wagon for one dollar. (Photo courtesy Zella Pelloux Slayton collection)

Bill Norton lived in a sheepwagon when he attended high school in Douglas, Wyoming. Norton restored sheepwagons for many years until his death in 2000. (Author's photo)

rancher obliged and Peyton lived in a sheepwagon for a school year, parked next to the one-room schoolhouse. At one point Peyton held classes for the five pupils in her wagon until a skunk that had set up his home in the schoolhouse was run off.[4]

Sometimes students lived in sheepwagons, too. When ranch kid Bill Norton attended a small-town Wyoming high school in the 1930s, his father moved a sheepwagon onto a vacant town lot that became young Billy's home for a year. Norton also remembered transient cowboys staying in his father's sheepwagons for a night or two before they moved on.[5]

Sheepwagons provided ideal temporary living quarters until a more permanent structure could be erected in boom towns like Salt Creek, near Casper, Wyoming. Later named Midwest after the Midwest Oil Company, Salt Creek began in 1910 as an oil camp. Early housing consisted of "some sheepwagons, tent houses, and about thirteen tarpapered shacks and several more shacks for bunkhouses."[6] Period photographs also show heavy equipment being freighted to Salt Creek with a cooster wagon bringing up the rear of the cargo.

A sheepwagon also served as conveyance for a long journey. A rancher in western South Dakota

Oilman Paul Stock, on far right, lived in a sheepwagon during an oil boom at Salt Creek, Wyoming. Stock, a self-made millionaire, later became the single largest private shareholder in Texaco Oil Company. (Photo courtesy Paul Stock Foundation, Cody, WY)

used a sheepwagon with four horses to travel with his family to Yellowstone Park in the early 1900s.[7] Another rancher in a remote area of Wyoming used one to take his wife to the nearest town for the birth of their child.[8]

Author Frances Seely Webb told of her 140-mile childhood trip in a sheepwagon from Casper to the hot springs at Thermopolis, Wyoming in 1897.[9] She made this journey with her sister and mother, another child, and his mother and grandfather who drove the four-horse team. They traveled twenty miles a day, stopping at streams to do laundry, and replenishing their supplies at road ranches or tiny towns along the way.

Adventures included the sheepwagon getting stuck in a stream and their rescue by two cowboys. Another time, a rock placed behind the wheels stopped the sheepwagon from rolling backwards while climbing a steep mountain pass. Helpful freighters showed them the proper way to position the wagon at night to withstand strong winds. The men simply dug a hole for each of the four wheels which helped to anchor the sheepwagon. When the party finally returned to Casper, Webb recalled being "happy and glad to be home after the somewhat crowded conditions of life in a sheepwagon."[10]

Sheepwagons were used by road crews during the 1930s. A photograph of a Civilian Conservation Corps crew at work on a roadbed features in the background the unmistakable shape of a sheepwagon, which appeared to be used as mobile storage for equipment.

During the Depression years when residential building came to a halt, people totally unrelated to the sheep business lived in sheepwagons. More than a few married couples can recall spending their first year of married life in a sheepwagon because there was nowhere else to live. If a rancher had a vacant sheepwagon, he would often allow a couple to set up housekeeping in it until they found a more permanent home. A woman remembered that rowdy friends from town visited their wagon one night and rocked it until the couple awakened and invited them in to continue the party.[11]

And finally, a couple was even married in a sheepwagon! The 1998 Wyoming Historical Calendar noted that in Casper, Wyoming, on January 4, 1901, the "1st wedding in sheepwagon solemnized."

This photo, circa 1935, raises more questions than it answers. Note the men in white shirts and ties to the right. Why are the two men loading (or unloading) sheep into the pickup? Why is the sheep-wagon inside the pen? Why is the wagon's door open? (Photo courtesy Wyoming Department of State Parks and Cultural Resources, Carrigan Collection)

The grace of the sheepwagon frame is evident in this photograph taken during renovation. (Photo courtesy Harve Nye)

CHAPTER FOURTEEN

Today's Sheepwagon Restorers

A COTTAGE INDUSTRY HAS developed in the West since the mid-1980s; individuals are buying old sheepwagons and renovating them for high resale value. The restored, sometimes totally rebuilt, wagons with gleaming, often deluxe interiors, have been removed from their original context and acquired a new function as a guest room, an office, a child's playroom, a decorative yard ornament, or an expensive piece of trendy western memorabilia for the wealthy celebrity.

Today's sheepwagon buyers often plan to use them as a getaway parked in their yard or at the summer cabin. Although few want to live in them full-time, what people are responding to is the charm and simplicity of life in the wagons, a feeling that author Ivan Doig expressed so beautifully when he described his 1940s childhood impression of the cozy house on wheels.

"Inside the wagon with Dad or McGrath, it felt to me as if space, the very air, had changed, somehow tidied and tamped itself. I wanted to live in a sheepwagon, so much more interesting were they than the

This sheepwagon was restored by the late John Burke, a prominent sheep rancher in Natrona County, Wyoming. As a young boy Burke spent time at the Schulte Hardware Company in Casper, watching the blacksmiths build and repair sheepwagons. He collected the old Schulte blacksmith tools and used them in his restorations. This wagon was sold for $7,000 in 1993 to a Vermont couple who used it as their summer cabin in the White Mountains. (Author's photo)

bland room back at the Camas. Nowhere else had the sense of deft shrinkage as if a house had been pulled in and pulled in until it came down just above your head and out past your fingers...the wagon had a Dutch door; with its bottom half closed, I could lean out on it and feel as if far up on a lookout across this high pasture of summer."[1]

Philosophies of sheepwagon restoration range from the traditional to the trendy, with middle-of-the-road compromises found in between. As in the past, depending on one's taste and pocketbook, sheepwagons can be customized to suit the individual. What all restorers share is a love of the unique vehicle, an innate understanding of the importance of its fading history in a time and place, the twenty-first-century American West, that is undergoing rapid change. They all express a sincere desire to preserve, in some form, the legacy of the sheepwagon.

Rawhide Johnson of Cody, Wyoming is a purist and traditionalist when it comes to wagon restoration. Examples of his exquisite and authentically restored chuck wagons, Yellowstone Stagecoaches, and sheepwagons are displayed in the world-famous Buffalo Bill Historical Center in Cody. He approaches his carefully selected old wagons much as a museum conservator

Rawhide Johnson of Cody, Wyoming is an expert wagon restorer. In addition to sheepwagons and chuck wagons, he has restored Yellowstone Stage Coaches. One of his restored sheepwagons is on display at the Buffalo Bill Historical Center in Cody, Wyoming. (Author's photo)

Sheepwagon restorer Jeff Laub transports an old wagon bought at an auction to his shop in Casper, Wyoming, 1999. (Author's photo)

Many of Terry Baird's restored wagons have hot and cold running water. (Photo courtesy Terry Baird collection)

would an eighteenth-century Philadelphia highboy carved by a superb craftsman. In fact, Rawhide resembles that museum professional in the philosophy they both share—leave as much as possible of the original material, including the paint, intact. A master of all components of wagon construction—iron, wood, and leatherwork—Rawhide shuns the belt sander and new bright paint so freely used by other wagon restorers. He believes the history and originality of a wagon can be told in the layers of paint and varnish he carefully removes by hand.

Rawhide's been working with sheepwagons for over thirty years. A ranch kid with roots in Wyoming, Montana, and Idaho, he displayed natural mechanical abilities by the time he was a teenager, when nearby ranchers brought their broken equipment, including sheepwagons, to him for repair. He admits to an evolution from purely practical repair to authentic craftsmanship in his approach to wagon restoration.[2]

Although Rawhide will customize according to a prospective buyer's wishes, it's clear that his heart, soul, and talent are in the authentic restoration of sheepwagons. Like most restorers, Rawhide finds his wagons in a variety of places; ranchers and auctions provide much of his raw material.

Terry Baird restored this wagon in 1993. The wagon sold for $35,000 to actress Nicole Kidman who purchased it as a Christmas gift for her then-husband, actor Tom Cruise. (Photo courtesy Terry Baird collection)

The interiors of Baird's wagons are often beautifully paneled and outfitted with cowboy collectibles and western antiques. (Photo courtesy Terry Baird collection)

Another type of sheepwagon restorer is Terry Baird of Big Timber, Montana, carpenter by trade, who works in partnership with interior designer Hilary Heminway. Their wagons represent the high end of 1990s cowboy chic, and have been featured in the glossy magazine *Architectural Digest* and at prestigious western design shows. Featuring such amenities as skylights, running water, interior vintage wood paneling, and cowboy collectibles, his wagons have fetched as much as $45,000 a pop. Luminaries such as Tom Cruise own a rehabilitated sheepwagon which Baird delivered to his Telluride, Colorado ranch.

Baird loves the texture, feel, and "pleasing decay" of the old wood in the sheepwagons he rescues. His goal is to turn a sow's ear into a silk purse by building "an environment with character, a place where people can feel good about themselves." He and Heminway accommodate their upscale clients by adding the non-traditional features they prefer.[3]

Wagon restorer Vern Krinke of Auburn, Washington recalls that his interest in sheep "camps" was sparked when he was thirteen years old and worked with a Basque herder. Krinke, now in his seventies, still remembers the cast-iron skillets, Dutch oven, "sourdough everything," and great strong coffee served

Pete and Sandy Roussan of Meeteetse, Wyoming (1998) stand in front of wagons which they plan to restore. They were among the first sheepwagon restorers in Wyoming. (Author's photo)

up in the wagon, and he vowed then that he would someday own one. He restored his first wagon in the 1970s, mounted on a Studebaker running gear he found in a junk pile.

Krinke's approach to sheepwagon restoration is a mix of the old and new. He might replace the rotted wood cabinet behind the stove with a new oak one. A stickler for detail, he always adds a bookshelf, rifle racks, an old coffee grinder, and the enamelware that was so common on the range. His wagons receive a gleaming new paint job in the traditional red and green Studebaker colors.

Krinke restores his wagons partly as a tribute to the historic sheep business and the sheep people who he believes played such an important role in western history. He takes great satisfaction in selling his wagons to people who he believes will preserve them, thereby keeping some piece of the history of the sheep business alive. A number of his wagons are in private carriage collections in New Hampshire and Oregon.

Restorer Harve Nye of Canyon City, Oregon prefers to make "absolutely historically correct and authentic" sheepwagons, modeled after those wagons manufactured in the early 1900s. A former boat builder, Nye appreciates the similarities between boats and the home on wheels: "They are both miniature spaces that must have everything a person needs to live in them for long periods." An ace wheelwright and expert woodworker, Nye uses traditional blacksmithing and joinery techniques in his work. His wagons sell from $17,000 to $25,000.[4]

"Before and after" photos of a wagon restored by Harve Nye. This wagon appeared in the movie The Ballad of Little Joe. (Photos courtesy Harve Nye)

C H A P T E R   F I F T E E N

The Meaning of Sheepwagons Today

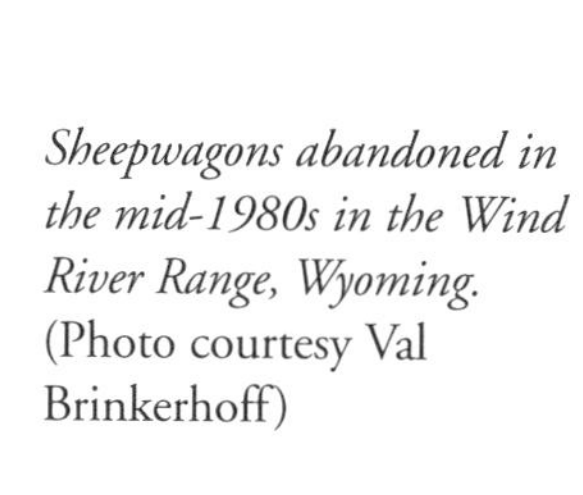

Sheepwagons abandoned in the mid-1980s in the Wind River Range, Wyoming. (Photo courtesy Val Brinkerhoff)

THAT WILSON CAMPS, now the sole commercial manufacturer of sheepwagons in the United States, makes only fifteen to twenty-five wagons a year, half for non-agricultural use, reflects the fact that the working sheepwagon is just about obsolete. As in other industries, the modernization of the business has, ironically, led to the discarding of those traditional tools once so crucial to its initial success—in this case the sheepwagon and herder which sustained the industry for so long. Other factors, of course, come into play.

The western sheep industry's fortunes have sunk to an all-time low, which the historical statistics reflect. The number of Wyoming stock sheep peaked in 1910 at 5,480,000; fell to 3,778,000 in 1940; 2,360,000 in 1960; and declined another 50% by 1980. The January 1, 2000 tally of just 460,000 stock sheep in the entire state is the industry's lowest count since before the 1880 total of 517,000.[1]

Obviously, the western sheep industry is threatened more today that at any other time in its 120-year

AHEAD
2—C
SEE ME
1909
SHEEP WAGON

Left: A sheepwagon "Welcome Wagon" greets visitors to Thermopolis, Wyoming, 1995. (Author's photo)

A sheepwagon is used to advertise a sporting goods store in Thermopolis, Wyoming. This is a new rather than restored wagon. Local rancher Dwight Lyman built this and other new wagons in the 1990s. (Author's photo)

A sheepwagon advertises a pizza shop in Casper, Wyoming, 1993. (Author's photo)

Pat Nagle and her grandson, Hunter Hatch, sit on the steps of her restored sheepwagon which she uses as a greenhouse, Casper, Wyoming, 1999. (Author's photo)

history. Faced with low prices for both wool and lamb, a weak predator-control program, loss of the wool incentive, and the hotly debated grazing reform, many third- and fourth-generation sheep ranchers are considering bailing out, as others who saw the writing on the wall have done over the past twenty years. When a rancher sells out, what also goes on the auction block are the implements and tools related to that outfit that often symbolize the entire industry. One of the most sought-after of those items today is the sheepwagon.

Although sheepwagons are a hot item for restorers, many wagons are beyond salvation; ranchers abandoned them at the spot where they came to rest many years ago. From various points along Interstate 80 in Wyoming, one can see three or four dilapidated sheepwagons on what was once a vibrant sheep landscape. The herder's wagon, for years the most recognizable symbol of the sheep business, is becoming, like the industry itself, an endangered species, a thing of the past.

But the sheepwagon is so unique and original that the image endures. Once scorned by those outside the industry, it is being transformed, like the cowboy before it, into a romantic symbol of the West. It seems the further the sheepwagon recedes into our past, the more popular it has become. The sheepwagon image pops up in many places: on a café placemat; as a Christmas ornament or mailbox; on postcards and notecards; as a logo for the Wyoming Wool Growers. Old sheepwagons are being recycled for advertising purposes, as novelty accommodations at dude ranches, in rodeo parades, as welcome wagons on the outskirts of small towns, or information booths at rodeos and state fairs.

Recently, hot on the heels of a resurgence of interest in cowboy verse, comes sheepherder poetry. As many a transplanted Easterner has discovered, one only has to talk the talk to qualify for the title—though it may be self-proclaimed—of "cowboy poet." Ditto for the sheepherder writings, although to be fair to both kinds of poets, their credibility must be earned by experiencing, up close and personal, the life they often humorously describe. Many of the herder poems have sheepwagons as their subject.

Sheepherder poetry has become popular in the past decade for the very same reasons that sheepwagons are being restored in record numbers. People are in search of a connection to the romanticized West of a bygone era, and the sheepherder and sheepwagon today are novel and accessible symbols of that past. The verse

The restored sheepwagon as a modern-day camper. The Steve Aagard family of Laramie, Wyoming uses their wagon for family camping trips in the Medicine Bow Mountains of southern Wyoming. (Photo courtesy Aagard family collection)

Near right: The entrance to the Camino sheep ranch near Buffalo, Wyoming uses the sheepwagon motif, 1999. (Author's Photo)

Far right: A sheep rancher who ran for political office pictured his sheepwagon on a deck of playing cards for a campaign promotional item, 1994. (Photo courtesy Richard Collier, Wyoming State Historic Preservation Office, Department of State Parks and Cultural Resources)

Sheepwagon mailboxes are becoming popular items. Just like the vehicle itself, no two are alike (Author's photos)

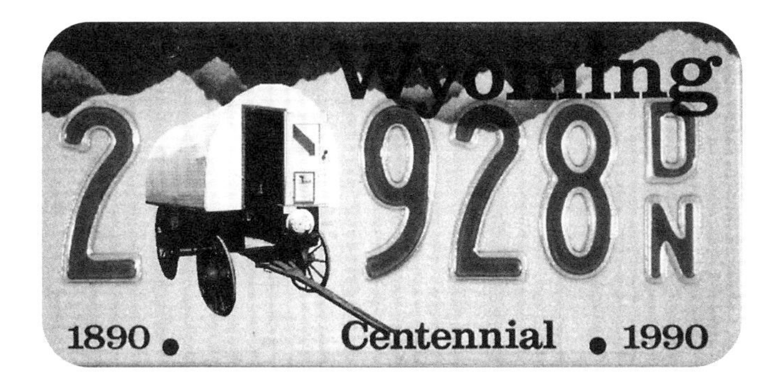

The sheepwagon was so important to Wyoming's economy that perhaps it should replace the bucking horse and cowboy on the state's license plate. (Images courtesy Carol Marander)

conjures up a lifestyle that in retrospect seems to be so much simpler than the world we live in today. Sheepherders and sheepwagons were once so closely connected to the earth's cycles and the care of defenseless animals that perhaps a man's work in that era strikes us as having a purer motivation than our own. Compare the herding life to contemporary employment, which for many takes place at a computer terminal within a small cubicle in a high-rise building with an artificially altered environment. Sheepherder poetry and sheepwagons both function as avenues to tap into a history we can never live, for it is gone.

The sheepwagon was for so long a common sight on the western landscape, and became so familiar, that it has been overlooked as an important part of the region's material culture. Entire books are devoted to artifacts of the cattle industry such as barbed wire, hats, boots, ropes, saddles; not so with the sheep business. Other than a few articles every ten or fifteen years about the quaint sheepwagon or the lonely life of the rapidly disappearing sheepherder, no definitive history of the western sheep business has been written since 1948.

Perhaps times are changing. In the year 2001, museums throughout the West are beginning to focus on the history of the sheep industry; many exhibits showcase a historic sheepwagon. Like an arrowhead that represents an earlier culture and relates something of the history of a time and place, perhaps the sheepwagon, as the most important cultural artifact and symbol of the historic western sheep industry, can serve a similar purpose, as a vehicle to tell the story.

E N D N O T E S

Part 1: The Story

Chapter 1. The Early Days of the Western Sheep Business

1. This poem by Alzada, Montana sheep rancher/poet E.F. Bischoff (1899-1984) was published in the *Wyoming Rural Electric News*, n.d. The author wishes to thank LaVonna Beardsley for bringing the poem, housed in the Warren Livestock Collection, to her attention.
2. Edward Norris Wentworth, *America's Sheep Trails: History: Personalities* (Ames, Iowa: Iowa State College Press, 1948), pp. 258, 270-271, 285.
3. United States Agriculture Department, National Agricultural Statistics Service (NASS). The author wishes to thank Richard Coulter, director of the Wyoming NASS, for his help with interpretation of the historical agricultural statistics.
4. Wallace Stegner, *Beyond the Hundredth Meridian: John Wesley Powell and the Second Opening of the West* (New York: Penguin Books, 1992), p. 214.

Part 2: The Wagon

Chapter 2. Inside the Basic Sheepwagon

1. Published in *Wyoming Rural Electric News*, September 1995. Used by permission of the Norton family.
2. Don Meike, interview by the author, 14 November 1993. Other ranchers have concurred on the primary function of the door.

Chapter 3. The Origins of the Sheepwagon

1. George Shumway, Edward Durell, and Howard C. Frey, *Conestoga Wagon 1750-1850: Freight Carrier for 100 Years of America's Westward Expansion* (York, Pennsylvania: Early American Industries Association, Inc. and George Shumway, 1964), pp. 1-2.
2. Richard Dunlop, *Wheels West: 1590-1900* (New York: Rand McNally & Company, 1977), p. 99.
3. Nick Eggenhofer, *Wagons, Mules and Men: How the Frontier Moved West* (New York: Hastings House Publishers, 1961), p. 123.
4. Frances C. Carrington, *My Army Life and the Fort Phil Kearny Massacre* (Philadelphia: J. B. Lippincott Company, 1910), pp. 181-182, 185. The author thanks Rawlins, Wyoming historian Rans Baker for directing her to this source.
5. Thomas Hardy, *Far from the Madding Crowd* (New York: The New American Library, Inc., 1960), pp. 19-21. A 1979 unpublished paper, "Home on the Range: Utah Sheep Camps" by Utah historian Tom Carter, directed the author to this quote and provided other very useful information.
6. C. H. Ward-Jackson and Denis E. Harvey, *The English Gypsy Caravan: Its Origins, Builders, Technology and Conservation* (New York: Drake Publishers, Inc., 1973), pp. 40-44.
7. Ibid., pp. 37-38.
8. Ibid., pp. 28, 39.
9. Kenneth Grahame, *The Wind in the Willows* (New York: Holt, Rinehart and Winston, 1980), p. 22. The author thanks Graham Jackson of Leicestershire, Great Britain for directing her to this source.

10. Robert Perry, *Sheep King: The Story of Robert Taylor* (Grand Island, Nebraska: Prairie Pioneer Press, 1986), pp. 19, 27. Taylor's sheepwagon was crowded: "Bill Hogg, my foreman, the herder, and myself live in it."

11. Neal Blair, "The sheepherder's castle," *Wyoming Wildlife*, April 1976, p. 29.

12. Charles M. Sypolt, "Keepers of the Rocky Mountain Flocks: A History of the Sheep Industry in Colorado, New Mexico, Utah and Wyoming to 1940" Ph.D. diss. Department of History, University of Wyoming, 1974, p.278

13. Elizabeth Arnold Stone, *Uinta County: Its Place in History* (Glendale, California: Arthur H. Clark Co., n.d.). Old West Museum, Cheyenne, Wyoming; Neal Blair Papers, Box 3, File 4.

14. Agnes Wright Spring, "Sheep Wagon Home on Wheels Originated in Wyoming," *Wyoming Stockman-Farmer*, December 1940, pp. 1-3; Edward Norris Wentworth, *America's Sheep Trails*, p. 405; *The Ranchers* (Alexandria, Virginia: Time-Life Books, 1977), pp. 404-405.

15. Clel Georgetta, *Golden Fleece in Nevada* (Reno: Venture Publishing Company, Ltd., 1972), pp. 79-80.

16. Although she is rarely cited, Agnes Wright Spring's 1940 article "Sheep Wagon Home on Wheels Originated in Wyoming," is obviously the source for the many subsequent articles written by various authors about sheepwagons. Articles on sheepwagons seem to appear cyclically, every ten years or so, in newspapers and magazines.

17. Larry Brown, "Beware the bite of a blue-haired dog," *Casper Star-Tribune*, 2 August 1998.

18. Mrs. Fred W. Dodge to Spring, 12 October 1940. Vertical File, Wyoming State Archives.

19. Mrs. Frank Ferris to Mr. Fred W. Marble, 16 May 1955. Vertical File, Wyoming State Archives.

20. *Douglas Budget*, 9 July 1886.

21. "A Venerable Pioneer of Early Wyoming Days," WPA File 348, Wyoming State Archives. The 1936 article was a collaboration between Georgia B. Kelley and eighty-year-old Frank George.

22. *Gillette News-Record*, 27 July 1967, p. 9. WPA File 348, Wyoming State Archives.

23. Dunlop, *Wheels West*, p. 99.

24. Georgetta, *Golden Fleece in Nevada*, p. 80; Wentworth, *The Ranchers*, p. 96.

25. Georgetta, *Golden Fleece in Nevada*, p. 80; Blair, "The sheepherder's castle," p. 30.

26. Coulter, Wyoming National Agricultural Statistics Service.

27. Ward-Jackson and Harvey, *The English Gypsy Caravan*, pp. 94-95.

28. James Julius Wilson, "Putting the West on Wheels," *Western Horseman* magazine, June 1991. Wilson was the grandson of "Cooster" Svendsen.

29. J. Tom Wall, *Life in the Shannon and Salt Creek Oil Field*, 1973, p. 4 (publisher unknown).

30. Edward Mark McGough, *Echoes from the West* (Glendo, Wyoming: High Plains Press, 1999), p. 70.

31. This was told to the author by a Wyoming man whose name she neglected to jot down!

Chapter 4. Blacksmiths, Builders, and Women's Work

1. *Douglas Budget*, 29 December 1899.
2. Bill Taliaferro, interview by the author, 2 February 1994.
3. Wyoming State Business Directory, 1904-05; 1908-09, (Denver: The Gazetteer Publishing Company), pp. 172, 29.
4. *Bill Barlow's Budget*, Anniversary Edition, 1907.
5. Walter "Spud" Murphy, interview by author, 2 July 1994.
6. Vashti Huff, interview by author, 17 October 1993.
7. Zella Pelloux Slayton, interview by author, 17 October 1993.
8. Vashti Huff interview by the author; 17 October 1993, the term "honeymoon wagon" appears to be yet another

regional variant as many people familiar with the sheep business have never heard of the term.

9. Grayce Esponda Miller, *Buffalo Bulletin*, April 1961.

CHAPTER 5. THE COMMERCIAL MANUFACTURE OF SHEEPWAGONS

1. Dunlop, *Wheels West*, p. 97; Studebaker National Museum Web site www.studebakermuseum.org/studestory.htm.
2. A letter to the author from Studebaker National Museum dated 13 May 2000 noted that the first Studebaker sheepwagon appeared in Catalog No. 143, 1899, and the last was in Catalog No. 804, 1913. According to the Studebaker National Museum Web page, Studebaker was the only manufacturer to successfully switch from horse-drawn to gasoline powered vehicles which they first produced in 1904. The company continued to produce wagons until 1920.
3. Studebaker Brothers Catalogs No. 215 (1902), p. 31; No. 604 (1912), pp. 41,55,94.
4. Dunlop, *Wheels West.*
5. Tom Lindmier, interview by the author, 19 March 1993; correspondence 31 May 2000. Lindmier has researched the history of the Rice Company and owns Rice wagon No. 6, mounted on the Minnesota-manufactured Winona running gear. Studebaker National Museum correspondence states there are no statistics on the number of sheep camp beds the company produced.
6. *The Wyoming State Business Directory, 1906-1907* (Denver: The Gazetteer Publishing Company, 1906), p. 613. The 1908-1909 Directory lists this company's address as Afton, Wyoming.
7. Harve Nye, interview by the author, 28 January 2001.
8. "A Modern Wagon Manufactory," The *Ogden Morning Examiner*, n.d.; A Brief History of Sidney Stevens Implement Co.; The Oregon Historical Society collection.
9. The information on the three Utah companies came from a variety of sources including Tom Carter, "Home on the Range," a 1979 unpublished paper. The late Blanton Owen kindly provided the author with a copy of his unpublished paper, "Sheep Camps: Rolling Vernacular Architecture," and additional material, as did John Cowan. Other sources include the Joseph Palen collection at the American Heritage Center at the University of Wyoming and the Cheyenne Frontier Days Old West Museum collection. An ad for the Olson Brothers Sheep Camps appeared in a 1980 *National Wool Growers* magazine.
10. Emer Wilson, interview by the author, 6 July 1994; Doyle Wilson, interviews with the author, 13 Feb 1995; 4 July 2000; Wilson Camps, Inc. brochure.

PART 3: THE HERDING LIFE

CHAPTER 6. TRADITIONAL SHEEPHERDERS

1. This poem was recited to the author by the late Bill Norton, 15 October 1993.
2. Robert A. Murray, "Of symbols and public perceptions: A new look at the cowboy myth and Wyoming's hardhat reality," *Wyoming History News*, January 1995.
3. As told to Eric Thane, *Out West* magazine, December 1953.
4. Maurice Kildare, *Out West* magazine, Fall 1966.
5. Bill Willcox, *Casper Star-Tribune*, 27 July 1989.
6. Wentworth, *America's Sheep Trails*, p. 402.
7. Archer B. Gilfillan, *Sheep: Life on the South Dakota Range* (St. Paul: Minnesota Historical Society Press, 1993 edition), p. 3.
8. Ibid., pp. 4-5.
9. Bill O'Neal, *Cattlemen vs. Sheepherders: Five Decades of Violence in the West, 1880 - 1920* (Austin: Eakin Press, 1989), p. 2.
10. Various authors have discussed the cowboy/sheepherder dichotomy, among them: W. Thalmann Kupper, "Sheepherder: A Character of the Early Southwest," *The Southwestern Sheep and Goat Raiser*, 1 Dec 1938, pp. 69-70:

American Heritage Center collection, University of Wyoming; George Watson Rollins, *The Struggle of the Cattleman, Sheepman and Settler For Control of Lands in Wyoming,* (New York: Arno Press, 1979), pp. 247-251; Gilfillan, *Sheep*, pp. 230-243; O'Neal, *Cattleman vs. Sheepherders*, pp. 2-6.

11. Wentworth, *America's Sheep Trails*, p. 402.
12. *Casper Star-Tribune*, 5 Nov 1980; *Wyoming Wool Grower* magazine, June 1990, p. 16.
13. Wentworth, *America's Sheep Trails*, p. 327.
14. Ivan Doig, *Dancing at the Rascal Fair* (New York: Atheneum Publishers), 1987, and *This House of Sky* (New York: Harcourt Brace Jovanovich, 1978) pp. 169-170.
15. Ann Esquibel Redman, interview by the author, 5 May 1993; Leonard Hay, interview by the author, 23 February 1994; Vern Vivion, interview by the author, 1994. Peg Arnold, "Wyoming's Hispanic Sheepherders," *Annals of Wyoming*, Winter 1997, pp. 29-35.
16. Kupper, "Sheepherder," *The Southwestern Sheep and Goat Raiser*, American Heritage Center collection, University of Wyoming.
17. Gretel Ehrlich, "The View from Burnt Mountain," Vertical File, Fulmer Library, Sheridan, Wyoming.
18. Michael Mathes, *Sheepherders: Men Alone* (Boston: Houghton Mifflin Company, 1975), p. 79.
19. Wentworth, *America's Sheep Trails*, p. 402.
20. Mark H. Brown and W. R. Felton, *Before Barbed Wire: L. A. Huffman, Photographer on Horseback* (New York: Henry Holt and Company, 1956), p. 78.
21. Gilfillan, *Sheep*, pp. 37-38.
22. Willcox, *Casper Star-Tribune*.
23. Bill Taliaferro, interview by the author, 2 February 1994.
24. Mathers, *Sheepherders: Men Alone*, pp. 8-9.
25. Ibid., p. 87.
26. Ibid., p. 92.
27. Ibid., p. 9.
28. Simon Iberlin & Madeline Harriet, interview by the author, 14 April 1995. Vern Vivion, interview by the author, 1994.
29. Ivan Thornton, interview by the author, 25 June 2000.

Chapter 7. Sheepherding Practices and Tools

1. Coulter, Wyoming National Agricultural Statistics Service.
2. Wentworth, *The Ranchers*, Time-Life series, p. 96.
3. Moroni A. Smith, *Herding And Handling Sheep On the Open Range In U.S.A.* (Salt Lake City: 1918), p. 3. American Heritage Center collection, University of Wyoming.
4. Smith, *Herding And Handling Sheep*, p. 53.
5. Ibid., pp. 54-55.
6. Ibid., p. 56.
7. Ibid., p. 57.
8. Thomas Seddon Taliaferro to Seddon Taliaferro, 1 October 1925; Thomas Seddon Taliaferro collection, American Heritage Center, University of Wyoming.
9. Ibid., p. 57.
10. Edgar and Turnell, *Lady of a Legend*, p. 91.
11. Ibid., p. 91.
12. Wentworth, *America's Sheep Trails*, p. 406.
13. *The Story of Shep: The Dog Who Was Ever Faithful*, Fort Benton, Montana: The River Press, author unknown, n.d.
14. John Niland, *A History of Sheep Raising in The Great Divide Basin of Wyoming: Personal Recollections on the End of an Era* (Cheyenne, Wyoming: Lagumo Corp., 1994), p. 118.
15. Ibid., p. 119.

Chapter 8. The Basque

1. William A. Douglass, "Lonely Lives Under the Big Sky," p. 33, Neal Blair collection, Cheyenne Old West Museum.

2. The source for much of the information on Basques in this chapter is noted Basque expert William A. Douglass, co-author with Jon Bilbao of *Amerikanuak: Basques in the New World*, (Reno: University of Nevada Press, 1975). Douglass has written extensively about the Basque in various books and articles.
3. Douglass, "Lonely Lives Under the Big Sky," p. 30.
4. Douglass and Bilbao, *Amerikanuak*, p. 233.
5. Douglass, "Lonely Lives Under the Big Sky," p. 38.
6. Dollie Iberlin, "The Basque Web: A story about the Basque people of Buffalo, Wyoming" (*The Buffalo Bulletin*, 1981), pp. 12-14.
7. Douglass and Bilbao, *Amerikanuak*, p. 335.
8. Iberlin, *The Basque Web*, p. 17.
9. J. Patrick Bieter, "Letemendi's Boarding House: A Basque Cultural Institution In Idaho," *Idaho's Yesterdays: Journal of Idaho Northwest History*, Spring 1993, pp. 7-8.
10. *The Basque Studies Program newsletter*, a publication of the University of Nevada, Reno, April 1992, p. 11.
11. Bieter, "Letemendi's Boarding House," p. 6.
12. Florence Urizaga Camino, interview by the author, 16 October 1993.
13. Leonard Hay interview by the author, 23 February 1994; Bill Taliaferro interview by the author, 2 February 1994.

Chapter 9. Women in Sheepwagons

1. Kathleen Hemry, *Kathleen's Book: An Album of Early Pioneer Wyoming in Word and Picture* (Casper, Wyoming: Mountain States Lithographing, 1989), pp. 73-76.
2. Ibid., pp. 38-39.
3. Edgar and Turnell, *Lady of a Legend*, p. 77.
4. Ibid., p. 87.
5. Ibid., pp. 52-53.
6. Ibid., p. 53.
7. Ibid., p. 91.
8. Ibid., p. 91.
9. Ibid., p. 91.
10. Ibid., pp. 112-113.
11. Ibid., p. 113.
12. Ibid., p. 114.
13. Ibid., p. 117.
14. Adelaide Hook Gilmore memoir, Park County (Wyoming) Historical Society Archives. The author thanks Park County historian Jeannie Cook for directing her to this collection.
15. Ibid., p. 9.
16. Ibid., p. 13.
17. Ibid., p. 14.
18. Ibid., p. 12.
19. Ibid., pp. 11-12.
20. The information on Louise Brown Turk is from a variety of sources including an interview by the author on 3 September 1994. Another source is an early draft of her memoir, published under the title *Sheep! Autobiography of Louise Turk, Woman Sheepherder* in 2001 by Pentland Press, North Carolina.
21. Louise Brown Turk, manuscript of *Sheep!*.
22. Louise Brown Turk, interview by the author, 3 September 1993.
23. Louise Brown Turk, interview by the author, 3 September 1993.
24. Hattie Tolman, interview by the author, 6 June 1998

Chapter 10. Sheepherder Superstitions, Customs, & Pastimes

1. Doig, *This House of Sky*, p. 172.
2. WPA File No. 1396, Wyoming State Archives.
3. Agnes Wright Spring, "'Home on the Range' Has Wheels,"

The Westerner, November 1947, p. 72; Vertical File, American Heritage Center collection.

4. The author received this poem from the late Dr. Joseph Palen whose collection now resides in the American Heritage Center, University of Wyoming.

Chapter 11. Sheepwagons in War

1. O'Neal, *Cattlemen vs. Sheepherders*, p. 16.
2. Wentworth, *America's Sheep Trails*, p. 523.
3. Wentworth, *America's Sheep Trails*, p. 525.
4. WPA File No.1280, Wyoming State Archives.
5. O'Neal, *Cattlemen vs. Sheepherders*. O'Neal's book describes these incidents in great detail. See Wentworth, *America's Sheep Trails*, pp. 527-528 for another account of the Diamondfield Jack case.
6. O'Neal, *Cattlemen vs. Sheepherders*, pp. 130-148; Wentworth, *America's Sheep Trails*, pp. 541-543 are the primary sources for this section.
7. Wentworth, *America's Sheep Trails*, p. 543.

Chapter 12. Modern-Day Sheepherders

1. Simon Iberlin, interview by the author, 12 July 1993. Numerous ranchers have confirmed this with the author.
2. Most of this chapter's information is from interviews conducted by the author: Larry Garro, Executive Director, Western Range Association, 7 February 1994; Oralia Mercado, Executive Director Mountain Plains Agricultural Service, 14 July 1996, 28 January 2000; LaVonna Beardsley, Executive Assistant, Warren Livestock Company, 1 June 2000. Brochure, Mountain Plains Agricultural Service; Various ranchers confirmed the information provided.

Part 4: The Legacy

Chapter 13. Other Uses of Sheepwagons

1. From the book *Sheepherder Poetry* by Obediah Brown, 1991.
2. Nuhn, Elizabeth, "Memories of an Oil Field," *Annals of Wyoming*, Volume 8, No. 1, Spring 1986.
3. Jessie A. Bryant, "Wagon was home to the herder," *Wyoming Rural Electric News*, June 1980.
4. Pauline Peyton, interview by the author, 14 November 1993. This story is also related in *Pages From Converse County's Past* (Douglas: Wyoming Pioneer Association, 1986), pp. 725-726.
5. Bill Norton, interview by the author, 15 October 1993.
6. J. Tom Wall, *Life in the Shannon and Salt Creek Oil Field*, 1973, p. 15 (publisher unknown).
7. Lawrence Brown, *Buffalo Commons Memoirs* (Bowman, ND: printed by Grapevine Press, 1995), p. 162.
8. Wyoming State Historic Preservation Office files, George and Betty Taylor interview, 8 January 1986.
9. Francis Seely Webb, "A Trip From Casper to Thermopolis," *Casper Star-Tribune*, 4 December 1966.
10. Ibid., p. 6.
11. The author has talked to numerous women who have related similar stories of living in a sheepwagon the first year of their married life.

Chapter 14. Today's Sheepwagon Restorers

1. Doig, *This House of Sky*, pp. 172-173.
2. Rawhide Johnson, interviews by the author, 26 September 1995; 14 May 1997.
3. Terry Baird, interviews by the author, 11 November 1994; 26 September 1995; 5 February 2001.
4. Harve Nye, interview by the author, 28 January 2001.

Chapter 15. The Meaning of Sheepwagons Today

1. Coulter, NASS Wyoming. "Stock sheep" consist of breeding sheep and lambs.

BIBLIOGRAPHY

Books

Brown, Lawrence. *Buffalo Commons Memoirs*. Bowman, North Dakota: Grapevine Press, 1995.

Brown, Mark and Felton, W. R. *Before Barbed Wire: L. A. Huffman, Photographer on Horseback*. New York: Henry Holt and Company, 1956.

Brown, Obediah. *Sheepherder Poetry*. 1991.

Carrington, Frances C. *My Army Life and the Fort Phil Kearney Massacre*. Philadelphia: J. B. Lippincott Company, 1910.

Doig, Ivan. *Dancing at the Rascal Fair*. New York: Atheneum Publishers, 1987.

__________ *This House of Sky*. New York: Harcourt Brace Jovanovich, 1978.

Douglas, William A. and Bilbao, Jon. *Amerikanuak: Basques In The New World*. Reno: University of Nevada Press, 1975.

Dunlap, Richard. *Wheels West: 1590-1900*. New York: Rand McNally & Company, 1977.

Edgar, Bob and Turnell, Jack. *Lady of a Legend*. Cody, Wyoming: Stockade Publishing, 1979.

Editors. *Pages From Converse County's Past*. Douglas, Wyoming: Pioneer Association, 1986.

Editors. *The Old West: The Ranchers*. Alexandria, Virginia: Time-Life Books, 1977.

Eggenhofer, Nick. *Wagons, Mules and Men: How the Frontier Moved West*. New York: Hastings House Publishers, 1961.

Ehrlich, Gretel. *The Solace Of Open Spaces*. New York: Penguin Press, 1985.

Georgetta, Clel. *Golden Fleece in Nevada*. Reno: Venture Publishing Company, Ltd., 1972.

Grahame, Kenneth. *The Wind in the Willows*. New York: Holt, Rinehart, and Winston, 1980.

Gilfillan, Archer B. *Sheep: Life on the South Dakota Range*. St. Paul: Minnesota Historical Society Press, 1993.

Hardy, Thomas. *Far From the Madding Crowd*. New York: The New American Library, Inc., 1960.

Hemry, Kathleen. *Kathleen's Book: An Album Of Early Pioneer Wyoming In Word and Picture*. Casper, Wyoming: Mountain States Lithographing, 1989.

Iberlin, Dollie. *The Basque Web: A story about the Basque people of Buffalo, Wyoming....* Buffalo, Wyoming: The Buffalo Bulletin, 1981.

Mathes, Michael. *Sheepherders: Men Alone*. Boston: Houghton Mifflin Company, 1975.

McGough, Edward Mark. *Echos from the West*. Glendo, Wyoming: High Plains Press, 1999.

Niland, John: *A History of Sheep Raising in The Great Divide Basin of Wyoming: Personal Recollections on the End of an Era*. Cheyenne, Wyoming: Lagumo Corp., 1994.

Perry, Robert. *Sheep King: The Story of Robert Taylor*. Grand Island, Nebraska: Prairie Pioneer Press, 1986.

O'Neal, Bill. *Cattlemen vs. Sheepherders: Five Decades of Violence in the West, 1880-1920.* Austin: Eakin Press, 1989.

Rollins, George Watson. *The Struggle of the Cattleman, Sheepman and Settler For Control of Lands in Wyoming.* New York: Arno Press, 1979.

Shumway, George, Durell, Edward and Frey, Howard C. *Conestoga Wagon 1750-1850: Freight Carrier for 100 Years of America's Westward Expansion.* York, Pennsylvania: Early American Industries Association, Inc. and George Shumway, 1964.

Smith, Moroni. *Herding And Handling Sheep On the Open Range in U.S.A.* Salt Lake City: 1918.

Stegner, Wallace. *Beyond the Hundredth Meridian: John Wesley Powell and the Second Opening of the West.* New York: Penguin Books, 1992.

Wall, J. Tom. *Life in the Shannon and Salt Creek Oil Field.* 1973.

Ward-Jackson, C. H. and Harvey, Denis E. *The English Gypsy Caravan: Its Origins, Builders, Technology and Conservation.* New York: Drake Publishers, 1973.

Wentworth, Edward Norris. *America's Sheep Trails: History: Personalities.* Ames, Iowa: Iowa State College Press, 1948.

Catalogs, Directories

Studebaker Brothers. Catalog No. 215, 1903.

__________ Catalog No. 604, 1912.

Wyoming State Business Directory, 1904-05. Denver: The Gazetteer Publishing Company.

__________, 1906-07.

__________, 1908-09.

Magazines, Newspapers, Unpublished Documents, Websites

"A Modern Wagon Manufactory: A Brief History of Sidney Stevens Implement Co." *The Ogden Morning Examiner*, n.d. The Oregon Historical Society, Portland, Oregon.

Arnold, Peg. "Wyoming's Hispanic Sheepherders." *Annals of Wyoming*, Winter 1997.

Baird, Terry. Oral Interviews 11 November 1994; 26 September 1995; 5 February 2001. Nancy Weidel collection, Cheyenne, Wyoming.

Beardsley, LaVonna. Oral Interview 1 June 2000. Nancy Weidel collection, Cheyenne, Wyoming

Bieter, J. Patrick. "Letemendi's Boarding House: A Basque Cultural Institution in Idaho." *Idaho's Yesterdays: Journal of Idaho Northwest History*, Spring 1993.

Blair, Neal. "The sheepherder's castle." *Wyoming Wildlife*, April 1976.

Brown, Larry. "Beware the bite of a blue-haired dog." *Casper Star-Tribune*, 2 August 1998.

Bryant, Jessie A. "Wagon was home to the herder." *Wyoming Rural Electric News*, June 1980.

Camino, Florence Urizaga. Oral Interview 16 October 1993. Nancy Weidel collection, Cheyenne, Wyoming.

Carter, Tom. "Home on the Range: Utah Sheep Camps." 1979.

Casper Star-Tribune, 5 November 1980.

Dodge, Mrs. Fred W. Correspondence with Agnes Wright Spring, 12 October 1940. Vertical File, Wyoming State Archives, Department of State Parks and Cultural Resources, Cheyenne, Wyoming.

Douglas Budget, 9 July 1886. 29 December 1899. Anniversary Edition, 1907.

Douglass, William A. "Lonely Lives Under the Big Sky." Neal Blair collection, Cheyenne Frontier Days Old West Museum.

Ehrlich, Gretel. "The View from Burnt Mountain." vertical file, Fulmer Public Library, Sheridan, Wyoming.

Ferris, Mrs. Frank. Correspondence with Mr. Fred W. Marble, 16 May 1955. Vertical File, Wyoming State Archives, Department of State Parks and Cultural Resources, Cheyenne, Wyoming.

Francovic, Annette Huff. Oral Interview, 17 October 1993. Nancy Weidel collection, Cheyenne, Wyoming.

Garro, Larry. Oral Interview 7 February 1994. Nancy Weidel collection, Cheyenne, Wyoming.

Gillette News-Record, 27 July 1967. WPA Files 348, Wyoming State Archives, Department of State Parks and Cultural Resources, Cheyenne, Wyoming.

Gilmore, Adelaide Hook Collection. Park County Historical Society Archives, Cody, Wyoming.

Hay, Leonard. Oral Interview, 23 February 1994. Nancy Weidel collection, Cheyenne, Wyoming.

Iberlin, Simon. Oral Interview, 12 July 1993. Nancy Weidel collection, Cheyenne, Wyoming.

Iberlin, Simon and Harriet, Madeline. Oral Interview, 14 April 1995. Nancy Weidel collection, Cheyenne, Wyoming.

Huff, Vashti Henderson. Oral Interview, 17 October 1993. Nancy Weidel collection, Cheyenne, Wyoming.

Johnson, Rawhide. Oral Interviews 26 September 1995; 14 May 1997. Nancy Weidel collection, Cheyenne, Wyoming.

Kelley, Georgia B. and George, Frank. "A Venerable Pioneer of Early Wyoming Days." WPA File 348, 1936, Wyoming State Archives, Department of State Parks and Cultural Resources, Cheyenne, Wyoming.

Kildare, Maurice. "Some Men Need It Lonely." *Old West Magazine*, Fall 1966.

Kupper, W. Thalmann. "Sheepherder: A Character of the Early Southwest." *The Southwestern Sheep and Goat Raiser*, 1 December 1938. American Heritage Center, Laramie, Wyoming.

Lindmier, Tom. Oral Interview, 19 March 1993. Nancy Weidel collection, Cheyenne, Wyoming.

Meike, Don. Oral Interview, 14 November 1993. Nancy Weidel collection, Cheyenne, Wyoming.

Mercado, Oralia. Oral Interviews 14 July 1996; 28 January 2000. Nancy Weidel collection, Cheyenne, Wyoming.

Miller, Grayce Esponda. *Buffalo Bulletin*, April 1961.

Murphy, Walter "Spud". Oral Interview, 2 July 1994. Nancy Weidel collection, Cheyenne, Wyoming.

Murray, Robert A. "Of symbols and public perceptions: A new look at the cowboy myth and Wyoming's hardhat reality." *Wyoming History News*, January 1995.

National Wool Growers Magazine. 1980.

Norton, Bill. Oral Interview, 15 October 1993. Nancy Weidel collection, Cheyenne, Wyoming.

Nuhn, Elizabeth. "Memories of an Oil Field." *Annals of Wyoming*, Spring 1986.

Nye, Harve. Oral Interview, 28 January 2001. Nancy Weidel collection, Cheyenne, Wyoming.

Owen, Blanton. "Sheep Camps: Rolling Vernacular Architecture." n.d.

Palen, Joseph Collection. American Heritage Center, Laramie, Wyoming.

Peyton, Pauline. Oral Interview, 14 November 1995. Nancy Weidel collection, Cheyenne, Wyoming.

Price, Jimmie Faye Henderson. Oral Interview, 6 June 1996. Nancy Weidel collection, Cheyenne, Wyoming.

Redman, Ann Esquibel. Oral Interview, 5 May 1993. Nancy Weidel collection, Cheyenne, Wyoming.

Slayton, Zella Pelloux. Oral Interview, 17 October 1993. Nancy Weidel collection, Cheyenne, Wyoming.

Spring, Agnes Wright. "Sheep Wagon Home on Wheels Originated in Wyoming." *Wyoming Stockman-Farmer*, December, 1940.

__________. "Home on the Range Has Wheels." *The Westerner*, November, 1947. vertical file, American Heritage Center, Laramie, Wyoming.

Studebaker National Museum correspondence with Nancy Weidel, 13 May 2000. Nancy Weidel collection, Cheyenne, Wyoming.

Studebaker National Museum Website: www:studebaker museum.org/studestory.htm

Sypolt, Charles M. "Keepers of the Rocky Mountain Flocks: A History of the Sheep Industry in Colorado, New Mexico, Utah and Wyoming to 1940." Doctoral Dissertation, Department of History, University of Wyoming, 1974.

Taliaferro, Bill. Oral Interview, 2 February 1994. Nancy Weidel collection, Cheyenne, Wyoming.

Taliaferro, Thomas Seddon. Correspondence to Thomas Seddon Taliaferro, Jr. 1 October 1925. Taliaferro Collection, American Heritage Center, Laramie, Wyoming.

Taylor, George and Betty. Interview 8 Jan 1986. Wyoming State Historic Preservation Office files, Department of State Parks and Cultural Resources, Cheyenne, Wyoming.

Thane, Eric. "Sheepherder: The West's Lonely Hero." *Out West Magazine*, December 1953.

The Basque Studies Program newsletter. University of Nevada, 1992.

"The Story of ...Shep: The Dog Who Was Ever Faithful." n.d. The River Press: Fort Benton, Montana.

Thornton, Ivan. Oral Interview, 25 June 2000. Nancy Weidel collection, Cheyenne, Wyoming.

Tolman, Hattie. Oral Interview, 6 June 1998. Nancy Weidel collection, Cheyenne, Wyoming.

Turk, Louise. Oral Interview 5 September 1994. Nancy Weidel collection, Cheyenne, Wyoming.

United States Agricultural Department, National Agricultural Statistics Service (NASS). Richard Coulter, Wyoming NASS Director, Cheyenne, Wyoming.

Vivion, Vern. Oral Interview 12 November 1993. Nancy Weidel collection, Cheyenne, Wyoming.

WPA files, No. 1280; No. 1396. Wyoming State Archives, Department of State Parks and Cultural Resources, Cheyenne, Wyoming.

Warren Livestock Company Collection, Cheyenne, Wyoming.

Webb, Francis Seely. "A Trip From Casper to Thermopolis." *Casper Star-Tribune*, 4 December 1966.

Weidel, Nancy. "The Wyoming Sheepwagon." *Annals of Wyoming*, Summer 1995.

Wilson, Doyle. Oral Interview 13 Feb 1995; 4 July 2000. Nancy Weidel collection, Cheyenne, Wyoming.

Willcox, Bill. "The lonely life of a sheepherder." *Casper Star-Tribune*, 27 July 1989.

Wilson, Emer. Oral Interview 6 July 1994. Nancy Weidel collection, Cheyenne, Wyoming.

Wilson, James Julius. "Putting the West on Wheels." *Western Horseman Magazine*, June 1991.

INDEX

A special hardcover limited edition of only 200 copies of this volume was printed simultaneously with the trade paperback edition. This special edition is bound in Persimmon Firenze Kivar 7 and embossed with gold foil. It is designed to be sold without a dust jacket.

The text of both the special edition and trade paperback edition is composed in twelve-point Adobe Garamond. Display type is Chaparrel by ITC. This book is printed on sixty-pound Centennial Offset, a recycled, acid-free paper by Bang Printing.